CSR, Sustainability, Ethics & Governance

Series Editors

Samuel O. Idowu, London, United Kingdom
René Schmidpeter, Cologne Business School, Germany

More information about this series at http://www.springer.com/series/11565

Gianna Moscardo • Pierre Benckendorff
Editors

Education for Sustainability in Tourism

A Handbook of Processes, Resources, and Strategies

Springer

Editors
Gianna Moscardo
College of Business
Law and Governance
James Cook University
Townsville
Queensland
Australia

Pierre Benckendorff
UQ Business School
The University of Queensland
St. Lucia
Queensland
Australia

ISSN 2196-7075 ISSN 2196-7083 (electronic)
CSR, Sustainability, Ethics & Governance
ISBN 978-3-662-47469-3 ISBN 978-3-662-47470-9 (eBook)
DOI 10.1007/978-3-662-47470-9

Library of Congress Control Number: 2015951782

Springer Heidelberg New York Dordrecht London

Printed on acid-free paper

Springer-Verlag GmbH Berlin Heidelberg is part of Springer Science+Business Media (www.springer.com)

Contents

List of Figures

List of Tables

List of Contributors

Sharee Bauld is a tourism professional with 17 years' experience predominately in South-east Asia as an independent tourism consultant, with a particular interest in tourism policy, investment, and private sector development. Her experience with tourism is broad and in-depth encompassing small-medium business and enterprise development, wide-ranging tourism policy, market research and strategy, branding and promotions, ecotourism development, tourism investment, and tourism value chains.

Pierre Benckendorff is an educator and social scientist at The University of Queensland, Australia. He has more than 10 years of experience in the development and delivery of undergraduate and postgraduate curricula in introductory tourism management, tourist behaviour, international tourism, tourism transportation, tourism technologies, tourism futures, and tourism analysis. His research interests include consumer behaviour, the impact of new technologies on tourism, tourism education, and tourism scholarship, and epistemology.

Anna Blackman specializes in the areas of business coaching effectiveness, Human Resource Management, Business Management, and Well-being. She is specifically interested in building capacity with regional and rural tourism businesses. She is currently employed as a Senior Lecturer for James Cook University in the School of Business.

Harald Buijtendijk has worked in the field of international tourism development for over 10 years. As lecturer, consultant, and social entrepreneur, he approaches industry challenges related to sustainability from multiple angles. Since August 2011, he works as a part-time lecturer at the department of International Tourism Management Studies of NHTV University, Breda, The Netherlands.

Dianne Dredge is Professor in the School of Culture and Global Studies, University of Aalborg, Denmark. She has 20 years practical experience as a tourism and environmental planner working for a range of public and private organizations. Her research interests focus on tourism planning, policy, governance, and knowledge dynamics between research, practice, and tourism education. She has a commitment to action research and knowledge co-creation from genuine shared dialogue between practitioners and academic researchers.

David Fennell teaches and researches in the Department of Tourism and Environment, Brock University, Canada. He has written widely on ecotourism, tourism ethics, and moral issues tied to the use of animals in tourism. He is Editor-in-Chief of the Journal of Ecotourism.

Mikell Gleason is Assistant Director for Program Development with the University of Georgia's Global Programs in Sustainability where she also oversees academic relations. Her Ph.D. is in anthropology and she served in the Peace Corps in Africa. Mikell has taught on virtually every different type of GPS program in the past 6 years.

Gayle Jennings is the Director of Research, Imagine Consulting Group International. Her research agenda focuses on practical and applied research for business and industry, research training and education, qualitative methodologies, and quality tourism experiences. Gayle is also an Adjunct Professor of Tourism Management, Department of Tourism, Sport and Hotel Management, Griffith University, Gold Coast Campus.

Ulrike Kachel is a lecturer in Tourism at Leeds Beckett University. She has a strong background in both information technology and the tourism industry. Based on this expertise, she is also the course/programme leader for the online Travel Business degrees at Leeds Beckett. Ulrike's research interests include sustainable tourism management and marketing, consumer behaviour, environmental practices and learning, online research, as well as higher education research.

Dagmar Lund-Durlacher is Head of the Department of Tourism and Service Management and Dean of the Undergraduate School at MODUL University Vienna. Her current research interests focus on Environmental Management Systems and Eco Certifications Schemes for the tourism industry, Green Consumer Behaviour, and Corporate Social Responsibility.

Kevin Lyons is an Associate Professor of Tourism and Management and Director of Postgraduate Studies in the Newcastle Business School at the University of Newcastle, Australia. His research focuses upon the intersections between volunteering, tourism, and community engagement, regional tourism planning and management, and transformational learning through travel.

Gianna Moscardo is a Professor in the College of Business, Law and Governance at James Cook University. Her research interests include evaluating tourism as a sustainable development strategy with an emphasis on understanding the relationships between characteristics of tourism development and dimensions of destination community well-being. She is also interested in understanding tourist behaviour and experiences and the effectiveness of tourist interpretation.

Laurie Murphy is an Associate Professor in the School of Business at James Cook University. Her research interests focus on improving tourism's contribution to regional communities with an emphasis on tourism marketing, including a focus on the backpacker market, destination image and choice, and more recently destination branding and tourist shopping villages. More recently, her research has evolved to look at tourism's contribution to community well-being in general and on the relationship between work, leisure, and family well-being.

Christian Schott is a Senior Lecturer in Victoria Business School at Victoria University of Wellington, New Zealand. His research interests span: sustainable tourism, tourism and climate change, teaching and learning, and youth travel and self-development. He is Vice Chair of the Tourism Education Futures Initiative (TEFI).

Stephen Schweinsberg is a lecturer in sustainable management in the UTS Business School. Stephen coordinates the core unit Integrating Business Perspectives, which is completed by all first year students in the Bachelor of Business degree. He also teaches in a range of research methods and other subjects across the Bachelor of Management and Master of Management degree programs. Stephen's current research interests are in the areas of Honours pedagogy in business education, the social impacts of coal seam gas development in Australian rural communities, and national park based tourism management.

Krystina Stoner is a PhD student in the Warnell School of Forestry and Natural Resources at the University of Georgia. Her research interests include the assessment of learning outcomes of education abroad, transformative learning experiences within an international context, and understanding the role of universities in fostering a globally aware student citizenry.

Lee Stoner is Senior Lecturer in the School of Sport and Exercise at Massey University in New Zealand where he specializes in exercise physiology and cardiovascular disease, but with an interest in the globalization and internationalization of international education.

Michael Tarrant is Josiah Meigs Distinguished Teaching Professor in the Warnell School of Forestry and Natural Resources at the University of Georgia, and director of the office of Global Programs in Sustainability, providing overseas programmes focusing on the theme of sustainable development throughout the South Pacific. His research interests include the learning outcomes of study abroad, global citizenship, the human dimensions of natural resources and international protected area and wilderness management.

Kristin Tessman is Assistant Director for Program Operations with the University of Georgia's Global Programs in Sustainability. She received her J.D. from UGA and manages the office administration and public relations. Kristin regularly teaches on programmes throughout the South Pacific.

Mylene van der Donk is lecturer and researcher at NHTV Breda University of Applied Sciences since 2008. The topics in which she is interested and lectures in are related to sustainable planning and development of tourism and recreation sites. Recent projects involve tourism policy assessments, analysing protected area visitor experiences and development of online sustainability training modules for entrepreneurs.

Stephen Wearing is an Associate Professor at the University of Technology, Sydney (UTS). He has conducted numerous projects and lectures worldwide and is the author of 13 books and over 100 articles dealing with issues surrounding leisure and sustainable tourism. His practical experience as an environmental and park planner at local, state, and international level have provided him with real world experiences that he brings to his teaching and research. His research interests include ecotourism, community based and volunteer tourism, environmentalism, sociology of leisure, and tourism and social sciences in protected area management.

Hilary Whitehouse is a researcher with the Centre for Research and Innovation in Sustainability Education (CRISE) and an educator with the College of Arts, Society and Education at James Cook University in Cairns, Queensland. She teaches the Master of Education (Sustainability) programme as well as science education and research education. Through CRISE, she curates a climate change education blog to help Australian teachers access the best resources and latest information. Hilary is an executive member of the Australian Association for Environmental Education, which advocates for education for sustainability.

Erica Wilson is Senior Lecturer in the School of Tourism and Hospitality Management at Southern Cross University. She has researched and taught in sustainable tourism for over 10 years, more recently focusing her attention on education for sustainability and critical pedagogy. Erica is currently a member on the Gondwana Rainforests of Australia World Heritage scientific committee. Erica's other publications are in the areas of gender and tourism, leisure constraints, and qualitative methods in tourism.

Chapter 1
The Importance of Education for Sustainability in Tourism

Gianna Moscardo

Abstract Growing public concern about environmental issues are contributing to the pressure on governments, businesses, groups and individuals to find solutions that move towards a more sustainable relationship with the planet. A key element in achieving the goals of sustainability is education. The decade 2005–2014 was declared the United Nations Decade of Education for Sustainable Development (UNDEFSD) in recognition of this close relationship between education and sustainability. Although this declaration focussed considerable attention on the concept of Education for Sustainability (EfS), discussions about the nature of the relationship between education and sustainability began before this declaration and are likely to continue long after 2014. Despite the close link between sustainability and tourism, there has been little systematic attention paid to education for sustainability in tourism. Two exceptions to this are the BEST Education Network and the Tourism Education Futures Initiative. This chapter introduces and provides an overview of EfS, links it to sustainability in tourism, reports on the outcomes of the BEST Education Network Think Tank in 2011, and provides a framework for understanding EfS in Tourism.

Keywords Education for sustainability • Tourism education • Tourism and sustainability

1.1 The Sustainability Imperative and the Evolution of Education for Sustainability

In 2009 the Harvard MBA graduating class developed the MBA Oath (MBA Oath Organisation, nd), which begins with the statements that "*As a business leader I recognize my role in society. My purpose is to lead people and manage resources to create value that no single individual can create alone. My decisions affect the well-being of individuals inside and outside my enterprise, today and tomorrow.*" The Oath goes on to list a series of promises related to ethical behavior, a concern

G. Moscardo (✉)
James Cook University, Townsville, QLD, Australia
e-mail: Gianna.moscardo@jcu.edu.au

G. Moscardo, P. Benckendorff (eds.), *Education for Sustainability in Tourism*, CSR, Sustainability, Ethics & Governance, DOI 10.1007/978-3-662-47470-9_1

for social justice and responsibility, and a commitment to sustainability. This kind of action is one type of response to what Dhiman (2012) refers to as the sustainability imperative. According to this argument, media coverage and public and government concern over events such as the Global Financial Crisis of 2008/2009 and the Deepwater Horizon Oil Spill, have contributed to the widespread perception that business owners and managers are selfish, greedy, untrustworthy and prepared to risk the well-being of others to achieve their own narrow agendas (Dhiman, 2012; Moscardo et al., 2013). These pressures combined with increasing public awareness of sustainability issues mean that businesses and governments are taking sustainability seriously (Dhiman, 2012; Esty & Winston, 2009; Moscardo et al., 2013).

The increased consciousness of sustainability can be linked to several United Nations global conferences beginning with the 1972 UN Conference on the Human Environment which established the United Nations Environmental Program and created a set of 26 principles for tackling sustainability. Principle 19 outlined the need for environmental education both within and outside formal schooling to "*broaden the basis for enlightened opinion and responsible conduct by individuals, enterprises and communities*" (UNEP, 1972). This principle was developed further through the 1977 Intergovernmental Conference on Environmental Education held in Tbilisi, Georgia. The Tbilisi Declaration established a framework for environmental education focused on addressing sustainability issues (UNESCO, 1977) and the importance of education as a tool for sustainability was reiterated in various statements made in the 1987 Brundtland report, Our Common Future (World Commission on Environment and Development, 1987). In the Foreword, Brundtland states that "*the changes in attitudes, in social values, and in aspirations that the report urges will depend on vast campaigns of education, debate and public participation.*" (p. 9).

As discussions of sustainability in general moved beyond an exclusively environmental focus, UN discussions of education moved from environmental education to education for sustainable development (EfSD). The 1992 UN Conference on Environment and Development, also known as the first Earth Summit, held in Rio De Janeiro, generated Agenda 21. This more detailed sustainability action plan devoted a whole chapter (36) to the importance of education, public awareness and training calling on national governments to implement national EfSD programs (United Nations, 1992). Concerns about the lack of progress in this aspect of sustainability prompted the development of a United Nations Decade focused on education for sustainable development commencing in 2005 (UNESCO, nd a).

There is ongoing debate about use of the word development in connection to sustainability with some arguing that development is an inappropriate concept to link to sustainability and preferring the use of just sustainability (Robinson, 2004). As might be expected, there has been some criticism of the UN use of education for sustainable development, rather than education for sustainability (McKeown, 2002). In practice the two terms are often used interchangeably (McKeown, 2002) and this chapter will use the EfS label except when specifically referring to

UN documents or statement. Despite criticism, this UN activity provides a detailed consideration of what EfS could or should be.

In this chapter EfS will be formally defined as "*the continual refinement of the knowledge and skills that lead to an informed citizenry that is committed to responsible individual and collaborative actions that will result in an ecologically sound, economically prosperous, and equitable society for present and future generations*" (President's Council on Sustainable Development, 1997). As noted previously, EfS evolved out of environmental education, extending into considerations of cultural, social and economic issues (Johnston & Johnston, 2013). Tilbury's (1995) discussion of this evolution noted that EfS also involved a move beyond education **about** environmental and sustainability issues to incorporate education **in** environments and communities and education **for** environments and communities. This distinction between education **for** sustainability and education **about** sustainability is an important one that is often confused. Education **about** sustainability focusses on knowledge of key issues such as climate change. Education **for** sustainability includes such knowledge but also incorporates a wider set of skills necessary to respond to these issues, explicitly examines values and ethics, and takes a larger holistic approach to sustainability (McKeown, 2002).

1.2 EfS: Objectives and General Principles

According to Cupitt and Smith (2012, p. 5) EfS should give learners "the ability to solve problems through knowledge integration, leading to big picture or world-view understandings with long term visions and an integration of values". In the formal educational context, Lake (2012) argues that the objectives for education in general are the same as those for EfS stating that "adequately preparing our students for contemporary life means that we need to ensure they are capable of applying critical and integrative thinking to complex situations, engaging in dialogue that is productive, and returning to and revising prior decisions ". Carp (2013), Blewitt (2013a) and Sterling (2013) also note the parallels in the objectives for education in general and EfS specifically and go on to suggest that current approaches to education do not achieve these objectives. It has even been argued that current educational systems, especially in higher education, actually contribute to and support unsustainable production and consumption systems (Blewitt, 2013a, 2013b; Jones, Selby, & Sterling, 2010a). Carp (2013, p. 223) suggests that "academic knowledge practices are complicit in creating, justifying, maintaining, and applying the behavior that places us at risk". Sterling (2013) goes further and argues that despite the addition of the label sustainability to educational policies, plans and curricula, current trends in education related to increasing competition and a market orientation towards students are taking the system away from, rather than towards, sustainability.

This gap between objectives and outcomes and concerns that current educational systems are part of the problem lead many to conclude that EfS requires

Table 1.1 Summary of key aspects of transformative education for sustainability

Existing approaches	Transformative approaches
Objectives	
Cognitive outcomes dominate	Cognitive, affective & skills outcomes in balance
Knowledge accumulation	Capacity building & self-development
Driven by short term needs of current economic system	Driven by long term human & environmental well-being
Extrinsic values	Intrinsic values
Teaching approaches	
Teacher centred	Learner centred
One/few teachers	Multiple participants learning together
Transmission of facts	Discovery of meaning
Single "objective" perspective	Multiple, relative, critical perspectives
Focus on theory	Focus on practice and action
Institutional systems	
Curriculum and structure organized by disciplines and content	Curriculum and structure organized by problems and skills
Intellectual analysis of issues	Analysis to support action
Focus on western models of education	Synthesis of different cultural approaches to learning
Ignore informal learning	Recognise, structure and use informal learning

Cotton and Winter (2010); O'Brien et al. (2013); Rowe and Johnston (2013); Ryan and Cotton (2013); Sterling (2004)

transformative education including both fundamental change in the nature of education overall (Cotton & Winter, 2010; Johnston & Johnston, 2013; O'Brien et al., 2013; Sterling, 2004, 2013; Tilbury, 2004, 2013; UNESCO, nd b; Wals, 2010) and a focus on education as a transformative experience for learners (Redman & Wiek, 2013; Rowe & Johnston, 2013). Table 1.1 summarises the key aspects of educational transformation proposed for EfS. These aspects indicate the need for a paradigm shift in education that requires that the whole educational process be redesigned. It is not sufficient to simply add units about sustainability into an existing program or curriculum (Cotton & Winter, 2010; Sterling, 2013).

These discussions of transformational education give rise to three general principles for EfS:

- EfS goes beyond the classroom;
- EfS requires whole system and/or whole institution change; and
- EfS must follow the core principles of sustainability.

The extension of learning beyond the classroom includes a number of dimensions. The first is a *stakeholder* dimension with explicit recognition of the need for public education campaigns, social marketing, training for decision makers in government and business, and support for community learning to reach people outside the formal educational system (Blewitt, 2013a, 2013b; O'Brien et al., 2013; Tilbury, 2004; UNESCO, nd b). The second is a *place* dimension with a focus on learning activities that are physically located outside classrooms and that involve interaction and collaboration with a diverse range of stakeholders (Cusick, 2012; Dawson, 2013; UNESCO, nd b). The third dimension is a *temporal* one and refers to the need to seriously address life-long learning. While many formal educational institutions often refer to life-long learning in their mission or vision statements, the focus is usually on encouraging a return to formal education or training driven by the demands of a dynamic global economic system (Blewitt, 2004). Blewitt (2004, 2013a) argues that life-long learning is better seen as an approach to everyday living based on a predisposition to think about and reflect critically on one's experiences and be willing to change. In this context life-long learning is something individuals do and the goal of the educational system is to encourage and develop this mindset.

Another way to think about learning beyond the classroom is to recognise the importance of informal learning and the existence of the hidden curriculum. The hidden curriculum refers to aspects of the educational system that convey information about what the system or institution values. The hidden curriculum includes the behaviors of educators and the policies and practices of the institutions (Ryan & Cotton, 2013). The second key theme in the transformation required for EfS is that attention must be paid not just to what and how things are taught but also to the hidden curriculum and how sustainability is embedded in the whole organisation, process or system. Some have also argued for a distinction between the taught and the learnt curriculum (Cuban, 1992; Glatthorn & Jailall, 2009). It can be argued that what is taught is not always what is learnt and what is learnt is not always what is taught.

Sterling (2004, 2013) extends this idea arguing for the creation of sustainable education (SE). He describes SE as helping to sustain people, communities and ecosystems, being ethically defensible, acting with integrity, justice, respect and inclusiveness, and being healthy and durable. This means addressing the environmental, social and economic impacts of the physical elements and other systems used in education. While there has been considerable discussion of ways to improve the environmental performance of educational institutions (cf. Hopkinson, James, & Van Winsum, 2004; Orr, 2010) much less attention has been given to the well-being of staff and students in formal education (cf. Sayce, Bradley, Ritson, & Quinn, 2013) and even less concern demonstrated for issues of access to, and inclusion in, education. Current formal educational systems, especially higher education, often reproduce and extend existing social inequalities and people are often excluded because of their ethnicity, gender, religion, socio-economic status and location (Ryan & Cotton, 2013; Sterling, 2013).

The UNESCO guidelines for EfSD (nd e) describe inclusion as more than just access to education. It is also about retention and educators need to address issues such as bullying and intolerance, as well as the provision of programs and physical spaces that support the health and comfort of all learners. Sterling (2013) provides a set of guidelines for achieving this whole institution approach in higher education and Eames, Barker, Wilson-Hill, and Law (2009) provide an example for primary and secondary schools.

While numerous barriers to implementing this kind of whole system change have been identified (cf. Blewitt, 2013b; Carp, 2013; Jones et al., 2010a; Sterling, 2013), one of the most important and entrenched is that of disciplinary boundaries. The increasing division of both education and knowledge into disciplines is seen as contributing to our inability to address the real-world problems that underpin sustainability (Johnston & Johnston, 2013). A focus on disciplines often leads to actions that serve to maintain the existence and boundaries of the disciplines, making it difficult to teach about complex problems, encouraging fragmented thinking, and limiting communication and innovation (Jones, Selby, & Sterling, 2010b). Moving beyond disciplines and teaching people to work in trans-disciplinary modes is both a major pre-condition and a challenge for EfS (Cotton & Winter, 2010).

1.3 Sustainability Values

The changes needed for EfS also require people to embrace and actively pursue a specific set of values (Cook, Cutting, & Summers, 2010). According to Coyne and Coyne (2001, p. 58) "*values are the glue that holds societies together*". Values can be defined as a type of belief that identifies "*desirable end states or modes of conduct . . . that transcends specific situations, [and] guides selection or evaluation of behavior, people, and events*" (Schwartz, 1994, p. 20). Values are the more abstract ideals that direct our attitudes, decisions and actions. The UN proposes freedom, equality, solidarity, tolerance, peace, respect for nature, and responsibility for others as key human values necessary for sustainability action (Torbjörnsson, Molin, & Karlberg, 2011; UNESCO, nd c). These are consistent with the results of research into the links between values and engaging in sustainability action. The available evidence, although limited, suggests that people are more likely to support sustainability programs and engage in sustainability behaviors if they place a higher importance on caring about the welfare of others, avoiding harm to others, harmony with nature, egalitarianism, a world at peace, and acting for the collective good (de Vries & Petersen, 2009; Held, Müller, Deutsch, Grzechnik, & Welzel, 2009; Kasser, 2011; Scholtens & Dam, 2007; Schultz et al., 2005; Shafer, Fukukawa, & Lee, 2007).

Knowing what values support sustainability is only part of the challenge for EfS, it is also necessary to highlight or activate these values. This activation of desirable values is variously referred to as civics, citizenship education, character education, moral education or values education, with the latter term recognised as the broadest and most commonly used label (Halstead & Taylor, 2000; Howard, Berkowitz, & Schaeffer, 2004; Veugelers, 2000). Values education is not a new approach in education but has become more important in recent years in response to rising concerns about various sustainability issues (Brady, 2011; Sharrock, 2010) and is explicitly included in many descriptions of EfS (Howard, 2012; Ryan & Cotton, 2013; UNESCO, nd d). Moscardo and Murphy (2011) provide a review of approaches to values education and UNESCO (nd c) provides an overview of teaching strategies that support values education. While these two sources provide details on a number of strategies that can be used in values education both conclude that a core element of values education is that the desired values are embedded throughout the curriculum and through all aspects of the institution and demonstrated by the educators and institutional leaders. This means that educators have to examine and reorient their own values (Cook et al., 2010; Cotton & Winter, 2010; Howard, 2012).

1.4 Skills and Knowledge to Support Sustainability Action

Given the trans-disciplinary, long-term, systemic, complex, and often contested nature of the various problems that need to be solved in order to move towards a more sustainable system of living, there is agreement that EfS curricula need to be designed around real world problems, especially those linked to local communities, and focused on skills and capacity building (Cotton & Winter, 2010; Frisk & Larson, 2011; Howard, 2012; Jones et al., 2010b). Table 1.2 summarizes the findings from several reviews of the competencies proposed for EFS curricula.

Although the primary emphasis in EfS is on skills and capacity building, there is also considerable consensus about the core content areas that need to be included in educational programs and these are mostly connected to the major sustainability issues (Clarke, 2012; Cotton & Winter, 2010; Cupitt & Smith, 2012). UNESCO offers two ways to consider this knowledge required to support sustainability action—a list of 11 sustainability issues that it argues require urgent attention (UNESCO, nd d) and a set of topics that are covered in its online teacher education program (UNESCO, nd c). Table 1.3 contains these two lists.

1.5 EfS: Teaching/Learning Strategies and Educational Philosophies

Given the emphasis on real-world problems and the types of skills listed in the previous section, it is not surprising to find that discussions about EfS teaching and learning strategies highlight approaches such as problem-based enquiry, experiential learning, service learning, collaborative learning, and the development of projects with local communities and stakeholders (Blewitt, 2004, 2013a; Cotton & Winter, 2010; Dawson, 2013; Frisk & Larson, 2011; Lake, 2012; Shrivastava, 2010). In addition, the use of case studies, field trips and experiments, role plays, simulations, debates, reflexive accounts and action research projects are suggested for EfS (Cotton & Winter, 2010; Frisk & Larson, 2011).

Table 1.2 Core skills in EfS

Core theme	Skills
Systems Analysis & Thinking	Understand interconnectedness of systems How to deal with complexity Working with uncertainty and making tradeoffs Connections between different scales of analysis
Futures Visioning	Future scenario building Forecasting & backcasting
Stakeholder interaction and engagement	Cross-cultural understanding, interaction and communication Conflict resolution Ability to take multiple/alternative perspectives on an issue
Action/change agency	Leadership Strategic planning Implementation Politics, governance and power Building & maintaining teams/networks/partnerships Effective persuasive communication
Meta-cognitive understanding	Self-reflection Ability to critically analyze assumptions and existing ideas Respect for evidence Awareness of bias and ability to distinguish between sound and unsound thinking Ability to synthesize and integrate knowledge
Innovation skills	Knowledge sharing Creativity

Frisk and Larson (2011); Morin (1999); O'Brien et al. (2013); Rowe and Johnston (2013); Wals (2010); Wiek, Withycombe, and Redman (2011)

Table 1.3 UNESCO EfS issues and topics

Key sustainability issues	Topics to be covered in EfS
Biodiversity Climate change Disaster risk reduction Cultural diversity Poverty reduction Gender inequality Health promotion Sustainable lifestyles Peace and human security Water conservation Sustainable Urbanisation	Exploring global realities • Examining major current global issues • Relationships between the issues Understanding sustainable development • Definitions of sustainability and sustainable development • History of sustainable development • Millennium development goals Understanding global patterns of production and consumption Culture and religion Indigenous knowledge and sustainability Women and sustainable development Population and development Understanding world hunger Sustainable agriculture Sustainable tourism Sustainable communities Globalisation Climate change Sustainability values, ethics and civic responsibility Sustainable futures

In discussion of EfS two particular educational approaches are noteworthy—critical pedagogy and place-based learning. Critical pedagogy has been given considerable coverage in the education literature (cf. Cho, 2010). What is relevant to the present discussion is that the broad aims of critical pedagogy are consistent with those outlined for EfS. Giroux (2007, p. 1) defines critical pedagogy as seeking to educate "students to become critical agents who actively question and negotiate the relationships between theory and practice, critical analysis and common sense, and learning and social change." Descriptions of critical pedagogy in action could offer frameworks for approaching EfS (cf. Kahn, 2010).

Place-based education or place conscious learning also offers a framework for designing learning experiences in EfS (cf. Somerville, 2010). This approach advocates engaging learners in activities that are embedded in and focussed on local places and spaces and using these activities to address global issues (Gruenewald & Smith, 2008). Cusick (2012) offers an example of this approach in a case study using ecotourism in Hawai'i to address a number of facets of sustainability. This raises the possibility of using tourism and tourist experiences as a tool for learning about sustainability with both holidaymakers and students.

1.6 BEST EN 2011 Think Tank on Education for Sustainability in Tourism

The previous sections have reviewed key aspects of EfS in general. In addition to this general approach there also exists literature exploring the implications of EfS for specific sectors. For example, the *Principles of Responsible Management Education* (PRME, nd) is a program designed to encourage business-related educators to adopt the principles of EfS. In tourism the BEST Education Network focusses on tourism education and sustainability. This network is an international collaboration between educators, mostly from higher education, who share an interest in improving education to support sustainable tourism. The group holds an annual Think Tank which uses a workshop format to develop research agendas and curriculum resources in a specific area relevant to tourism and sustainability. In 2011 the Think Tank was hosted by Temple University in Philadelphia and the topic was education for sustainability in tourism (BEST EN, 2011).

The 2011 Think Tank used a Nominal Group Technique (Stewart, Shamdasani, & Rook, 2007) to address the question: "What content should be included and/or what issues need to be addressed in university tourism curricula in order to move towards EfS in tourism?" Participants were asked to write down as many ideas as possible in a 10 min time period with each idea written on a separate post-it note. After all the ideas were written participants were invited to place their post-it notes on a white board. After the first few participants had placed their ideas on the white board, subsequent participants were asked to add their ideas to those already on the board to create themed clusters of ideas. Tables 1.4 and 1.5 contain a summary of

Table 1.4 Main themes for values, knowledge, skills/capabilities for EfS in tourism

Themes	Summary of ideas/issues
Sustainability values	
Values	Values that support sustainable tourism Discussion of values Recognition that it requires changes in values/attitudes Need to understand values connected with generations/cultures Eastern/western philosophies that shape sustainability approaches
Ethics	Examine own personal ethics Sustainability ethics
Knowledge	
Knowledge to support sustainable tourism	Links between travel and tourism and carbon impacts and climate change Sustainability is not just environment—its society, culture and economy Challenges of defining sustainable tourism Principles of sustainability—there is no such thing as sustainable tourism Can mass tourism be sustainable tourism? Impacts of tourism Differences in awareness of sustainable tourism products and services
Persuasive communication	Modes/media of communication Models of effective persuasive communication How to identify key target audiences and what you know about them
Stakeholders	Who are they? What motivates them? Different stakeholder perspectives Collaborative approaches Current good practices Importance of partnerships
Interconnectivity	Tourism as a vehicle for intercultural understanding Tourism and its relationship to Quality of life and destination well-being Looking beyond tourism to understand connections to sustainability in other areas Sustainable livelihoods approach Supply chain for sustainable tourism Links between Corporate Social Responsibility and sustainability
Future directions	What might the issues be in the future Need to stay in touch with new developments The future of tourism is at risk
Skills/capabilities	
General	How to educate others such as employees, government officials Principles of interpretation for tourists

(continued)

Table 1.4 (continued)

Themes	Summary of ideas/issues
	Leadership Community engagement Innovation Politics of sustainability discourse
Reflection	Self as part of a bigger picture Learners need to reflect on their role in the social engineering of sustainable tourism Need to acknowledge their own values
Dealing with complexity	Don't get overwhelmed by the problem Respective roles & responsibilities of all stakeholders in achieving sustainability How to deal with complexity Sustainable tourism is about tradeoffs Concepts and boundaries are blurring There are often no right or wrong answers Sustainable tourism is a multi-dimensional concept

Table 1.5 Main themes for teaching/learning strategies and challenges for EfS in tourism

Themes	Summary of ideas/issues
Teaching/learning strategies	
Tools for teaching Sustainability	Role playing Debates Learn by doing Experiential learning outside the classroom Practice what you preach Sustainability principles modelled in the classroom (e.g., paperless, diverse, links to communities) Simulations Case studies
Other	
Support for Educators	Need to understand which methods are effective for educating visitors/communities Need to understand the cognitive models that underpin learning/perceptions Need to understand the conditions that support sustainability learning
Challenges	Difficult to assess outcomes Need to live sustainability, not just teach it Need to get beyond students to other stakeholders How can tourism be used to enhance tourist and community knowledge of sustainability Not the main concern of industry Does generate revenue for low resource communities

the results of this exercise organized according to the main themes that were identified in each aspect of EfS.

Examination of this summary indicates considerable convergence with the broader EfS literature with two gaps, the need for whole system change and trans-disciplinarity. The former is likely to reflect the focus of the workshop on university level tourism and hospitality programs and an awareness of the limited power of educators in this context to make whole system changes. The latter is likely to reflect that tourism as an area of academic interest already operates in at least an inter-disciplinary space, which further strengthens the argument for using tourism as a tool to support wider EfS.

More detailed examination of the written comments revealed considerable variation in reference to sustainability and tourism with some participants talking about tourism and sustainability, some about sustainability alone and others using the phrase sustainable tourism. In the latter case sometimes sustainable tourism appeared to be used in the broad sense of changing or improving all aspects of tourism to improve sustainability, that is, tourism becoming more closely aligned with sustainability. In others it seemed that sustainable tourism was a specific type of tourism or alternative to some assumed standard, often referred to as mass, tourism. This tension between the idea of all tourism becoming more sustainable and the idea of developing another form of tourism labelled as sustainable is an ongoing one in the wider tourism literature (Hall, 2011; Saarinen, 2006). In the present discussion it could be a major issue in EfS in tourism as the conception of sustainable tourism as a specific type of tourism is more likely to be addressed by adding a specific teaching unit about this alternative tourism rather than addressing the skills and capacity needed for improving sustainability more widely.

1.7 The Current Status of Education for Sustainability in Tourism

The UNESCO online teacher training package includes tourism and agriculture in its list of topics to be incorporated into EfS curricula (nd c). It seems that these two sectors are included because of their strong linkages to globalization, the extent and diversity of their potential impacts, both positive and negative, on all dimensions of sustainability, and that both are key targets for development agencies around the world. The inclusion of tourism in discussions of sustainability reflects a long history of concerns about the linkages between tourism and sustainability (Saarinen, 2013). Despite extensive discussions of tourism and sustainability over the last 30 years, there has been only limited attention paid to the relationships between education, tourism and sustainability with the BEST EN 2011 Think Tank the first event that focused exclusively on EfS and tourism.

The available relevant literature on tourism education can be considered according to the major stakeholders of formal educational institutions and students, tourists, destination communities, tourism businesses and other organisations. Discussions of tourism education in formal educational institutions have been dominated by debates about the balance between professional or vocational requirements and liberal education and subsequent suggestions for what should be included in curricula within university levels programs in tourism, hospitality and event management (Dredge et al., 2012). There has been very little discussion of tourism education in the vocational sector, except for some listings of what is or should be included in programs (cf. Fidgeon, 2010) and extensive literature searches for this chapter failed to find any detailed consideration of tourism in school education beyond vocational training in secondary level school systems. Further, the overwhelming majority of tourism education papers examined assumed the discussion was about tourism, hospitality and event courses and degrees. Whilst tourism specific units may occasionally appear in other courses and degrees, there has been very little analysis of how tourism is or could be addressed in non-tourism programs.

Research into what is included in tourism, hospitality and events programs in higher education consistently report that while sustainability as topic for tourism education is much discussed, it is not well-integrated into existing programs. Fidgeon's (2010) review of UK tourism and related curricula seems to suggest that many programs have incorporated sustainability and sustainable tourism elements as a response to perceived market demand rather than as part of any integrated and/or holistic approach to EfS. Day et al. (2012) after a detailed examination of tourism and related curricula in Australian higher education reported that many programs include units on sustainable tourism and/or tourism impacts, but few programs include other aspects of sustainability such as ethics or corporate social responsibility. Deale, Nichols and Jacque's (2009, p. 40) review of hospitality programs in the US concluded that "*what is striking is that it did not yet appear that SE [sustainability education] was integral to education in this field*".

Fidgeon (2010) did provide two examples of programs that have pursued more critical approaches to tourism, although these were not overtly linked to EfS. There have also been some examples of whole tourism curricula guided by EfS (cf. Millar et al., 2012) and several descriptions of EfS activities in teaching units within tourism degrees (cf. Jamal, Taillon, & Dredge, 2011; Jennings, Kensbock, & Kachel, 2010). Additionally there are suggested frameworks for thinking about EfS in tourism (cf. Canziani, Sönmez, Hsieh, & Byrd, 2012) and research into barriers to adopting EfS in tourism (cf. Wilson & von der Heidt, 2013). A common link in this emerging literature on sustainability in higher education tourism curricula is the Tourism Education Futures Initiative (TEFI, nd). TEFI is a network of tourism educators who meet annually to discuss challenges, creativity and change in tourism curricula in higher education. The initiative does not specifically target sustainability and tourism education but the meetings have generated a set of five values-based principles for tourism education which overlap considerably with the major competencies listed for EfS. These "TEFI Values" are:

- *Stewardship*: including service to the community, sustainability and responsibility;
- *Ethics*: including honesty, transparency and authenticity;
- *Mutuality*: including diversity, inclusiveness, equity, humility and collaboration;

- *Knowledge*: including critical thinking, innovation, creativity and networking; and
- *Professionalism*: including leadership, relevance, timeliness, reflexivity, teamwork and pro-activity (Sheldon, Fesenmaier, & Tribe, 2011).

The emergence of TEFI and the activities of BEST EN are indicative of the growing recognition of the need to address sustainability education in tourism, although the discussion to date in both organisations has been focused on tourism programs and higher education.

The second stakeholder group that can be considered are tourists. Falk, Ballantyne, Packer, and Benckendorff (2012) argue that the links between travel and learning have been relatively neglected by researchers despite the long history of travel as an educational activity and the recent increased interest in educational tourism. They do acknowledge the work on interpretation which has examined in detail the ways in which visitors use and respond to the explanation and information provided at tourist sites through such media as signs, displays, exhibitions, guide books, and guided tours, but note that research has tended to focus on learning of knowledge about the specific visited sites. Recent critical reviews of both research and practice in interpretation conclude that there is a need to better understand what tourists learn beyond factual knowledge of sites (Ablett & Dyer, 2009) and that there is an opportunity in tourism to engage in social marketing linked to sustainability (Ballantyne & Packer, 2011; Moscardo, 2015). In other words, tourist settings and travel experiences could be used for effective place-based EfS (Blewitt, 2013a).

Consideration of sustainability and education in tourism beyond students in higher education and tourists has been very limited. Discussions of tourism education for destination communities, tourism businesses, tourism staff and other tourism stakeholders have been almost exclusively about training in hospitality and tourism management to support tourism development (Moscardo, 2008). This focus on training for tourism rather than education about tourism limits the capacity of these stakeholders to effectively engage in tourism decision making and/or tourism business activities and can be linked to unsustainable tourism (Moscardo, 2011).

1.8 A Framework for Education for Sustainability in Tourism

Given the growing recognition of the importance of EfS, and evidence that there are educators and researchers within tourism beginning to re-examine the nature of tourism education, it seems timely to focus attention on EfS in tourism. Figure 1.1 provides a descriptive framework for thinking about the dimensions and aspects of EfS in tourism developed from both the earlier review of EfS and the BEST EN workshop outcomes. At the centre of the figure is the *individual learner* and this

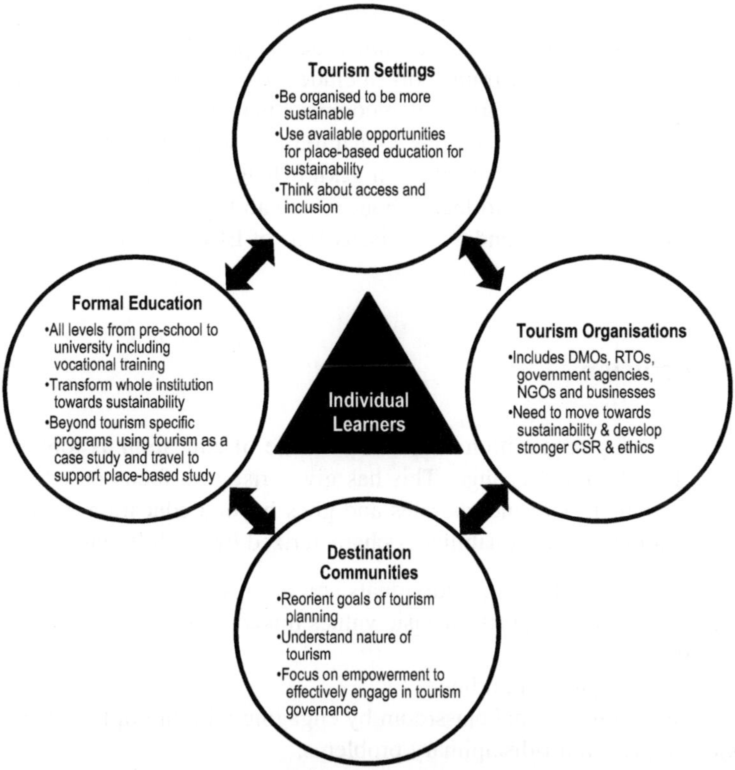

Fig. 1.1 Framework for education for sustainability (EfS) in tourism

includes students enrolled in formal educational programs at all levels, business owners, staff in tourism organisations, government decision makers, community members and tourists. The key *stakeholders* in EfS in tourism are identified as formal educational institutions, destination communities and tourism organisations. The development of effective EfS in tourism requires that all these stakeholders are engaged both as learners and educators. In this framework, for example, formal education institutions should seek opportunities both to assist destination communities to learn about tourism and sustainability and also to work with destination communities to create sustainability learning activities for individual learners at all levels. Similarly, tourism organizations should provide EfS programs for business owners and operators and other destination stakeholders, develop partnerships with formal educational institutions to improve skills and capacity for themselves in the area of sustainability in tourism, and work with other stakeholders to develop EfS opportunities for tourists at their destination.

While all the principles, values, skills, and learning strategies and approaches described for EfS in the earlier section apply to each of these stakeholders, the figure identifies some challenges for each of the three core stakeholder groups. In the case of formal educational institutions, for example, the discussion of tourism, education and sustainability has to be expanded beyond tourism specific programs in higher education and more holistic integrated approaches to sustainability need to be adopted. Finally, the framework includes tourism settings. This element is included for two reasons. Firstly, it recognises that the ultimate goal of EfS in tourism is change in tourism settings so that tourism supports sustainability both at and beyond the destination. Secondly, it acknowledges that these settings offer an opportunity for individuals to learn about sustainability. This framework offers a way to investigate, review and guide discussions of EfS in tourism.

1.9 Conclusion

There is a growing recognition of the importance of education as tool to support sustainability action and change. This has given rise to a focus on education **for** sustainability which both incorporates and goes beyond education **about** sustainability. Education **for** sustainability is characterized by the following features:

- It is transformative both for individuals and groups;
- It is driven by and supports intrinsic values linked to human and environmental well-being;
- It is aimed at capacity building;
- It goes beyond the formal classroom by engaging with multiple stakeholders to tackle complex multi-disciplinary problems;
- It takes a whole system approach; and
- It follows the core principles of sustainability.

Tourism and sustainability are closely linked and discussions about sustainability have a long history amongst tourism academics and practitioners. Current approaches to tourism education are, however, far removed from these EfS principles. With a few exceptions, there has been very little attention paid to sustainability and tourism education. This chapter has highlighted some of these exceptions and using the principles of EfS in general suggested a way to begin to create education for sustainability in tourism.

References

Ablett, P. G., & Dyer, P. K. (2009). Heritage and hermeneutics: Towards a broader interpretation of interpretation. *Current Issues in Tourism, 12*(3), 209–233.

Ballantyne, R., & Packer, J. (2011). Using tourism free-choice learning experiences to promote environmentally sustainable behaviour: The role of post-visit 'action resources'. *Environmental Education Research, 17*(2), 201–215.

BEST EN. (2011). Who we are. http://www.besteducationnetwork.org/aboutus.php. Accessed 21 Nov 2013.

Blewitt, J. (2004). Sustainability and lifelong learning. In J. Blewitt & C. Cullingford (Eds.), *The sustainability curriculum* (pp. 24–42). London: Earthscan.

Blewitt, J. (2013a). *The ecology of learning: Sustainability, lifelong learning and everyday life*. Hoboken, NJ: Taylor & Francis.

Blewitt, J. (2013b). EfS: Contesting the market model of higher education. In S. Sterling, L. Maxey, & H. Luna (Eds.), *The sustainable university* (pp. 51–70). Abingdon: Routledge.

Brady, L. (2011). Teacher values and relationship: Factors in values education. *Australian Journal of Teacher Education, 36*(2), 5.

Canziani, B. F., Sönmez, S., Hsieh, Y., & Byrd, E. T. (2012). A learning theory framework for sustainability education in tourism. *Journal of Teaching in Travel & Tourism, 12*(1), 3–20.

Carp, R. M. (2013). Towards a resilient academy. In L. F. Johnston (Ed.), *Higher education for sustainability* (pp. 223–237). New York: Routledge.

Cho, S. (2010). Politics of critical pedagogy and new social movements. *Educational Philosophy and Theory, 42*(3), 310–325.

Clarke, P. (2012). *Education for sustainability: Becoming naturally smart*. Hoboken, NJ: Taylor & Francis.

Cook, R., Cutting, R., & Summers, D. (2010). If sustainability needs new values, whose values? Initial teacher training and the transition to sustainability. In P. Jones, D. Selby, & S. Sterling (Eds.), *Sustainable education* (pp. 313–327). Hoboken, NJ: Taylor & Francis.

Cotton, D., & Winter, J. (2010). 'It's not just bits of paper and light bulbs': A review of sustainability pedagogies and their potential for use in higher education. In P. Jones, D. Selby, & S. Sterling (Eds.), *Sustainability education: Perspectives and practice across higher education* (pp. 39–54). Abingdon: Earthscan.

Coyne, K., & Coyne, R. (2001). Dispelling the myths of character education. *Principal Leadership, 2*(3), 58–61.

Cuban, L. (1992). Curriculum stability and change. In P. W. Jackson (Ed.), *Handbook of research on curriculum: A project of the American Educational Research Association* (pp. 216–247). New York: MacMillan.

Cupitt, J., & Smith, S. (2012). *Teaching for sustainability*. Melbourne: Macmillan.

Cusick, J. (2012). Teaching ecotourism in the backyard of Waikiki, Hawai'i. In K. A. Bartels & K. E. Parker (Eds.), *Teaching sustainability teaching sustainably* (pp. 140–160). Sterling, VA: Stylus.

Dawson, J. (2013). Ecovillages and the transformation of values. In *Worldwatch state of the world 2010* (pp. 185–190). New York: W.W. Norton.

Day, M., Walo, M., Weeks, P., Dredge, D., Benckendorff, P., Gross, M. J., et al. (2012). Analysis of Australian tourism, hospitality and events undergraduate education programs. Office of Learning and Teaching. www.olt.gov.au/.../CG9_1020_Dredge_2012_Analysis_THE_undergraduate_education.pdf. Accessed 12 Nov 2013.

de Vries, B. J. M., & Petersen, A. C. (2009). Conceptualizing sustainable development: An assessment methodology connecting values, knowledge, worldviews and scenarios. *Ecological Economics, 68*, 1006–1019.

Deale, C., Nichols, J., & Jacques, P. (2009). A descriptive study of sustainability education in the hospitality curriculum. *Journal of Hospitality & Tourism Education, 21*(4), 34–42.

Dhiman, S. (2012). Social responsibility: Caring for people, products, peace, preservation, and planet. In J. Marques, S. Dhiman, & S. Holt (Eds.), *Business administration education* (pp. 19–42). New York: Macmillan.

Dredge, D., Benckendorff, P., Day, M., Gross, M. J., Walo, M., Weeks, P., et al. (2012). The philosophic practitioner and the curriculum space. *Annals of Tourism Research, 39*(4), 2154–2176.

Eames, C., Barker, M., Wilson-Hill, F., & Law, B. (2009). Investigating the relationship between whole-school learning approaches to education for sustainability and student learning. Teaching and Learning Research Institute. http://www.tlri.org.nz/sites/default/files/projects/9245_summaryreport.pdf. Accessed 14 May 2011.

Esty, D., & Winston, A. (2009). *Green to gold: How smart companies use environmental strategy to innovate, create value, and build competitive advantage*. Hoboken, NJ: Wiley.

Falk, J. H., Ballantyne, R., Packer, J., & Benckendorff, P. (2012). Travel and learning: A neglected tourism research area. *Annals of Tourism Research, 39*(2), 908–927.

Fidgeon, P. R. (2010). Tourism education and curriculum design: A time for consolidation and review? *Tourism Management, 31*(6), 699–723.

Frisk, E., & Larson, K. L. (2011). Educating for sustainability: Competencies and practices for transformative action. *Journal of Sustainability Education, 2*(March), 1–20. http://www.jsedimensions.org/wordpress/wp-content/uploads/2011/03/FriskLarson2011.pdf. Accessed 20 Sept 2013.

Giroux, H. A. (2007). Introduction: Democracy, education, and the politics of critical pedagogy. In P. McLaren & J. L. Kincheloe (Eds.), *Critical pedagogy: Where are we now?* (pp. 1–8). New York: Peter Lang.

Glatthorn, A. A., & Jailall, J. M. (Eds.). (2009). *The principal as curriculum leader: Shaping what is taught and tested*. Thousand Oaks, CA: Corwin Press.

Gruenewald, D., & Smith, G. (2008). Introduction: Making room for the local. In D. Gruenewald & G. Smith (Eds.), *Place-based education in the global age* (pp. xiii–xxiii). New York: Lawrence Erlbaum.

Hall, C. M. (2011). Policy learning and policy failure in sustainable tourism governance: From first-and second-order to third-order change? *Journal of Sustainable Tourism, 19*(4-5), 649–671.

Halstead, J. M., & Taylor, M. J. (2000). Learning and teaching about values: A review of recent research. *Cambridge Journal of Education, 30*(2), 169–202.

Held, M., Müller, J., Deutsch, F., Grzechnik, E., & Welzel, C. (2009). Value structure and dimensions. Empirical evidence from the German world values survey. *World Values Research, 2*, 55–76.

Hopkinson, P., James, P., & Van Winsum, A. (2004). Learning by doing: Environmental performance improvement in UK higher education. In J. Blewitt & C. Cullingford (Eds.), *The sustainability curriculum* (pp. 78–91). London: Earthscan.

Howard, P. (2012). Who will teach the teachers? Reorienting teacher education for the values of sustainability. In K. A. Bartels & K. E. Parker (Eds.), *Teaching sustainability teaching sustainably* (pp. 149–157). Sterling, VA: Stylus.

Howard, R. W., Berkowitz, M. W., & Schaeffer, E. F. (2004). Politics of character education. *Educational Policy, 18*(1), 188–215.

Jamal, T., Taillon, J., & Dredge, D. (2011). Sustainable tourism pedagogy and academic-community collaboration: A progressive service-learning approach. *Tourism and Hospitality Research, 11*(2), 133–147.

Jennings, G., Kensbock, S., & Kachel, U. (2010). Enhancing 'Education About and For Sustainability'in a tourism studies enterprise management course: An action research approach. *Journal of Teaching in Travel & Tourism, 10*(2), 163–191.

Johnston, D. D., & Johnston, L. F. (2013). Introduction: What's required to take EfS to the next level? In L. F. Johnston (Ed.), *Higher education for sustainability* (pp. 1–8). New York: Routledge.

Jones, P., Selby, D., & Sterling, S. (2010a). Introduction. In P. Jones, D. Selby, & S. Sterling (Eds.), *Sustainable education* (pp. 1–16). Hoboken, NJ: Taylor & Francis.

Jones, P., Selby, D., & Sterling, S. (2010b). More than the sum of their parts? Interdisciplinarity and sustainability. In P. Jones, D. Selby, & S. Sterling (Eds.), *Sustainable education* (pp. 17–37). Hoboken, NJ: Taylor & Francis.

Kahn, R. (2010). *Critical pedagogy, ecoliteracy, & planetary crisis: The ecopedagogy movement*. New York: Peter Lang.

Kasser, T. (2011). Cultural values and the well-being of future generations: A cross-national study. *Journal of Cross-Cultural Psychology, 42*(2), 206–215.

Lake, D. (2012). Sustainability as a core issue in diversity and critical thinking education. In K. A. Bartels & K. E. Parker (Eds.), *Teaching sustainability teaching sustainably* (pp. 31–40). Sterling, VA: Stylus.

MBA Oath Organization. (n.d.). MBA oath: Responsible value creation. http://mbaoath.org/. Accessed 21 Nov 2013.

McKeown, R. (2002). ESD Toolkit. http://www.esdtoolkit.org/discussion/default.htm. Accessed 21 Sept 2013.

Millar, M., Brown, C., Carruthers, C., Jones, T., Kim, Y.-S., Raab, C., et al. (2012). Implementing environmental sustainability in the global hospitality, tourism and leisure industries. In L. F. Johnston (Ed.), *Higher education for sustainability* (pp. 124–136). New York: Routledge.

Morin, E. (1999). *Seven complex lessons in education for the future*. Paris: UNESCO.

Moscardo, G. (2008). Community Capacity Building—An Emerging Challenge for Tourism Development. In G. Moscardo (Ed.), *Building community capacity for tourism* (pp. 1–15). Wallingford: CABI.

Moscardo, G. (2011). The role of knowledge in good governance for tourism. In E. Laws, H. Richins, J. Agrusa, & N. Scott (Eds.), *Tourist destination governance* (pp. 67–80). Wallingford: CABI.

Moscardo, G., (2015). Stories of people and places: Interpretation, tourism and sustainability. In Hall, C. M., Gossling, S., & Scott, D. (Eds.). (2015). *The Routledge handbook of tourism and sustainability* (pp. 294–304). London: Routledge.

Moscardo, G., Lamberton, G., Wells, G., Fallon, W., Lawn, P., Rowe, A., et al. (2013). *Sustainability in Australian business: Principles and practice*. Brisbane: Wiley-Blackwell.

Moscardo, G., & Murphy, L. (2011). Toward values education in tourism: The challenge of measuring the values. *Journal of Teaching in Travel & Tourism, 11*(1), 76–93.

O'Brien, K., Reams, J., Caspari, A., Dugmore, A., Faghihimani, M., Fazey, I., et al. (2013). You say you want a revolution? Transforming education and capacity building in response to global change. *Environmental Science & Policy, 28*, 48–59.

Orr, D. W. (2010). What is higher education for now? In *Worldwatch state of the world 2010* (pp. 75–82). New York: W.W. Norton.

President's Council on Sustainable Development. (1997). From classroom to community and beyond: Education for a sustainable future. http://clinton2.nara.gov/PCSD/Publications/TF_Reports/linkage-top.html#toc. Accessed 10 May 2011.

PRME. (n.d.). PRME_principles for responsible management education. http://www.unprme.org/. Accessed 10 May 2011.

Redman, C. L., & Wiek, A. (2013). Sustainability as a transformation in education. In L. F. Johnston (Ed.), *Higher education for sustainability* (pp. 214–223). New York: Routledge.

Robinson, J. (2004). Squaring the circle? Some thoughts on the idea of sustainable development. *Ecological Economics, 48*(4), 369–384.

Rowe, D., & Johnston, L. F. (2013). Learning outcomes: An international comparison of countries and declarations. In L. F. Johnston (Ed.), *Higher education for sustainability* (pp. 45–59). New York: Routledge.

Ryan, A., & Cotton, D. (2013). Times of change: Shifting pedagogy and curricula for future sustainability. In S. Sterling, L. Maxey, & H. Luna (Eds.), *The sustainable university* (pp. 151–167). Abingdon: Routledge.

Saarinen, J. (2006). Traditions of sustainability in tourism studies. *Annals of Tourism Research, 33* (4), 1121–1140.

Saarinen, J. (2013). Critical sustainability: Setting the limits to growth and responsibility in tourism. *Sustainability, 6*(1), 1–17.

Sayce, S., Bradley, J. F., Ritson, J., & Quinn, F. (2013). Well-being: What does it mean for the sustainable university? In S. Sterling, L. Maxey, & H. Luna (Eds.), *The sustainable university* (pp. 211–232). Abingdon: Routledge.

Scholtens, B., & Dam, L. (2007). Cultural values and international differences in business ethics. *Journal of Business Ethics, 75*(3), 273–284.

Schultz, P. W., Gouveia, V. V., Cameron, L. D., Tankha, G., Schmuck, P., & Franěk, M. (2005). Values and their relationship to environmental concern and conservation behavior. *Journal of Cross-Cultural Psychology, 36*(4), 457–475.

Schwartz, S. H. (1994). Are there universal aspects in the structure and contents of human values? *Journal of Social Issues, 50*(4), 19–45.

Shafer, W. E., Fukukawa, K., & Lee, G. M. (2007). Values and the perceived importance of ethics and social responsibility: The US versus China. *Journal of Business Ethics, 70*(3), 265–284.

Sharrock, G. (2010). Two Hippocratic oaths for higher education. *Journal of Higher Education Policy and Management, 32*(4), 365–377.

Sheldon, P. J., Fesenmaier, D. R., & Tribe, J. (2011). The tourism education futures initiative (TEFI): Activating change in tourism education. *Journal of Teaching in Travel & Tourism, 11* (1), 2–23.

Shrivastava, P. (2010). Pedagogy of passion for sustainability. *Academy of Management Learning & Education, 9*(3), 443–455.

Somerville, M. J. (2010). A place pedagogy for 'global contemporaneity'. *Educational Philosophy and Theory, 42*(3), 326–344.

Sterling, S. (2004). An analysis of the development of sustainability education internationally: Evolution, interpretation and transformative potential. In J. Blewitt & C. Cullingford (Eds.), *The sustainability curriculum* (pp. 43–62). London: Earthscan.

Sterling, S. (2013). The sustainable university: Challenge and response. In S. Sterling, L. Maxey, & H. Luna (Eds.), *The sustainable university* (pp. 17–50). Abingdon: Routledge.

Stewart, D. W., Shamdasani, P. N., & Rook, D. W. (2007). *Focus groups: Theory and practice* (2nd ed.). Thousand Oaks, CA: Sage.

TEFI. (n.d.). About TEFI. http://www.tourismeducationfutures.org/about-tefi. Accessed 21 Nov 2013.

Tilbury, D. (1995). Environmental education for sustainability: Defining the new focus of environmental education in the 1990s. *Environmental Education Research, 1*(2), 195–212.

Tilbury, D. (2004). Rising to the challenge: Education for sustainability in Australia. *Australian Journal of Environmental Education, 20*(2), 103–114.

Tilbury, D. (2013). Another world is desirable: A global rebooting of higher education for sustainable development. In S. Sterling, L. Maxey, & H. Luna (Eds.), *The sustainable university* (pp. 71–85). Abingdon: Routledge.

Torbjörnsson, T., Molin, L., & Karlberg, M. (2011). Measuring attitudes towards three values that underlie sustainable development. *Utbildning och Demokrati, 20*(1), 97–121.

UNEP. (1972). Declaration of the United National Conference on the Human Environment. http://www.unep.org/Documents.Multilingual/Default.asp?documentid=97&articleid=1503. Accessed 21 Sept 2013.

UNESCO. (1977). Intergovernmental conference on environmental education. http://unesdoc.unesco.org/images/0003/000327/032763eo.pdf. Accessed 21 Sept 2013.

UNESCO. (n.d. a). Education for sustainable development. http://en.unesco.org/themes/education-sustainable-development. Accessed 18 May 2011.

UNESCO. (n.d. b). Four thrusts of ESD. http://www.unesco.org/new/en/education/themes/leading-the-international-agenda/education-for-sustainable-development/education-for-sustainable-development/four-thrusts-of-esd/. Accessed 18 May 2011.

UNESCO. (n.d. c). Teaching and learning for a sustainable future. http://www.unesco.org/education/tlsf/. Accessed 21 Sept 2013.

UNESCO. (n.d. d). Education for sustainable development—Leading the international agenda. http://www.unesco.org/new/en/education/themes/leading-the-international-agenda/education-for-sustainable-development/. Accessed 18 May 2011.

UNESCO. (n.d. e). Contributing to a more sustainable future. http://unesdoc.unesco.org/images/0014/001410/141019e.pdf. Accessed 21 Sept 2013.

United Nations. (1992). United Nations sustainable development. http://sustainabledevelopment.un.org/content/documents/Agenda21.pdf. Accessed 21 Sept 2013.

Veugelers, W. (2000). Different ways of teaching values. *Educational Review, 52*(1), 37–47.
Wals, A. E. (2010). Mirroring, gestaltswitching and transformative social learning: Stepping stones for developing sustainability competence. *International Journal of Sustainability in Higher Education, 11*(4), 380–390.
Wiek, A., Withycombe, L., & Redman, C. L. (2011). Key competencies in sustainability: A reference framework for academic program development. *Sustainability Science, 6*(2), 203–218.
Wilson, E., & von der Heidt, T. (2013). Business as usual? barriers to education for sustainability in the tourism curriculum. *Journal of Teaching in Travel & Tourism, 13*(2), 130–147.
World Commission on Environment and Development. (1987). *Our common future*. Oxford: Oxford University Press.

Part I
Tourism and Sustainability: Core Knowledge

Chapter 2
Tourism and Sustainability: Challenges, Conflict and Core Knowledge

Gianna Moscardo

Abstract There are a number of critical connections between tourism and sustainability as recognised in the consistent inclusion of tourism in UN discussions of sustainability and evident in the substantial government and academic literature on tourism and sustainability. This chapter has two main goals. First, it will provide an overview of the concept of sustainability and its connections to tourism for educators seeking to introduce it into a tourism education program. Second, the chapter will present a critical review of the concept of sustainable tourism as a way of encouraging a deeper understanding of the relationship between tourism and sustainability.

Keywords Sustainability • Sustainable tourism • Quality of life • Ethics • Tourism planning

2.1 Introduction

There is a long history of connection between tourism and sustainability organised around three related themes. The first is the widespread and longstanding use of tourism as a strategy for economic development in many regions, but especially in the regions sometimes referred to as part of the 'Global South' (i.e. Africa, Central and Latin America, and much of Asia) (Moscardo, 2008a). This makes tourism a focus of discussions about the effectiveness of different development options in achieving sustainability. The second theme is that tourism is a form of consumption characterised by the global North-South economic interactions that are often associated with sustainability problems (Redclift, 2005). Many environmental and social justice problems can be linked to situations where consumers from the Global North (i.e. North America, Western Europe and developed parts of East Asia) are able to use their discretionary income to buy goods and services produced cheaply in countries of the Global South (Wells, 2013). The lower costs of production in this Global South are typically based on lower wages, poor working conditions, lesser

G. Moscardo (✉)
James Cook University, Townsville, QLD, Australia
e-mail: Gianna.moscardo@jcu.edu.au

G. Moscardo, P. Benckendorff (eds.), *Education for Sustainability in Tourism*, CSR, Sustainability, Ethics & Governance, DOI 10.1007/978-3-662-47470-9_2

environmental regulation and greater negative environmental impacts (Desjardins, 2007). While not all tourism fits this model, there are sufficient examples of this situation to raise concerns about the global sustainability of tourism (Klein-Vielhauer, 2009). Finally, tourism has been described as both a contributor to and a victim of various phenomena linked to sustainability especially climate change (Scott, Hall, & Gossling, 2012).

Despite these connections, the considerable growth in the sustainability objectives and statements in tourism policies and plans, and the extensive academic discussion of "sustainable tourism", it is not clear that tourism practice has responded effectively to the challenges of sustainability. Many commentators have been critical of the way sustainability is conceptualised and implemented in tourism. This chapter will review the wider literature on what sustainability and sustainable development might and should mean. This review will include a brief history of the concepts of sustainability and sustainable development and provide an overview of the connections between sustainability and the related concepts of ethics, corporate social responsibility, governance and quality of life. The chapter will then critically examine the phenomenon of sustainable tourism and conclude by outlining a set of sustainability issues that have yet to be fully considered in tourism.

2.2 The Rise of Sustainability

A number of studies into media coverage and corporate reporting have documented the rise of sustainability as a topic of interest. Work by Barkemeyer, Figge, Holt, and Wettstein (2009), for example, analysing coverage of sustainability and related concepts in 115 national newspapers across the world, found steady incremental growth from 1990 to 2008 with sustainability becoming a well-established topic in the public arena in recent years. Similar analyses of corporate reporting reveal the same pattern of increasing attention paid to aspects of sustainability (Kolk, 2008). Costanza's (2009) discussion contrasting the empty and full world model of the global economy summarises the key issues that are driving this rise of sustainability. The existing or empty world model assumes that the natural environment has abundant supplies of resources for human use and ample capacity to absorb the waste produced by this human activity. This empty world model assumes the goal of the economy is the production of goods and services and that issues of social well-being are irrelevant. Costanza (2009) argues that this model no longer works as the massive growth and globalisation of the economy means that the negative impacts of human production and consumption are significant and costly, that there is increasing recognition of the finite limits to our use of the natural environment and that economic growth has not addressed issues of poverty, and that the current system does not include the cost of usage of and damage to public goods. He argues for a shift towards a full world model of the economy that recognises the importance of the natural environment and social issues and reaffirms the goal of the economy as the improvement of well-being or quality of life (QoL).

Table 2.1 A simple sustainability timeline

Phases	Events
Early concerns about the consequences of changing economies	1776 Smith writes about the negative consequences of free markets 1798 Malthus argues for limits to population growth 1892 Muir establishes the Sierra Club, a conservation organisation
Rising awareness of global environmental and social issues	1962 Carson's book, the Silent Spring, highlights major environmental problems arising from production systems 1965 First Fairtrade organization appears 1968 Hardin writes The Tragedy of the Commons describing the consequences of pursuing individual welfare at the expense of communal good
The United Nations begins discussing issues of poverty, justice and environmental damage	1972 UN Conference in Stockholm sets up the UN Environment Programme (UNEP) 1987 Publication of Our common Future, aka The Brundtland Report 1990 First report on climate change from the Intergovernmental Panel on Climate Change (IPCC) 1992 UN Earth Summit held in Rio de Janeiro 1993 Agenda 21 published 1993 First meeting of the UN Commission on Sustainable Development
Global business organisations begin to engage	1995 World Business Council for Sustainable Development (WBCSD) is established 1999 Launch of the Dow Jones Sustainability Index
The New Millennium	2002 UN Millennium Summit in Johannesburg sets goals for significant poverty reduction by 2015 2002 Global Reporting Initiative (GRI) releases first guidelines for business monitoring and reporting on sustainability performance 2006 Stern report on Climate Change 2007 Al Gore and the IPCC win a Nobel Peace Prize for work on identifying and communicating climate change issues

Adapted from Lumley and Armstrong (2004); Mebratu (1998); Paul (2008); Spindler (2013); Wells (2013)

Although the concepts of sustainability and sustainable development have become a central part of public discussion and debate in the last decade, the issues described by Costanza (2009) have much longer histories. Table 2.1 provides a simple timeline highlighting both the history of some of the key issues driving concerns about sustainability and some of the key international events related to sustainability in the last 30 years.

2.3 Sustainability and Sustainable Development: Complex and Contested Concepts

Despite this lengthy history there is considerable confusion and debate about what the terms sustainability and sustainable development actually mean (Scott, 2012). While it would be easy to use one of the most quoted definitions (Lozano, 2008) from the Brundtland Report (1987), which defined sustainable development as "*development that meets the needs of the present without compromising the ability of future generations to meet their own needs*" (p. 43), this definition has been the topic of much conflict and confusion. Detailed reviews of definitions and descriptions of sustainability and sustainable development consistently identify eight key themes—the Triple Bottom Line (TBL), well-being/QoL, the idea of multiple forms of capital, strong versus weak approaches to sustainability, the importance of ethics, justice and equity, the central role of corporate social responsibility (CSR); the value of sustainability as an integrative guiding ideal; and the need for transformative change (Antrop, 2006; Jabareen, 2008; Moscardo, 2013a; Parker, 2012; Redclift, 2005).

The first references to the Triple Bottom Line (TBL) concept are usually attributed to Elkington who described it as an idea that "*focuses corporations not just on the economic value that they add, but also on the environmental and social value that they add—or destroy*" (2004, p. 3). The concept was meant to direct attention to the need to consider three main aspects of business activity—environmental impacts, often referred to as planet, social impacts, or people, and economic impacts, or profits. The concept was subsequently further developed within accounting as a new way to measure and monitor business activity with the assumption that these new procedures would lead to changing practices (Lamberton, 2005). Recent analyses have, however, been very critical of the value of the concept for achieving any significant improvements in sustainability (Milne & Gray, 2013). Despite these criticisms the idea that sustainability has three dimensions—environmental, economic and social—continues to be a pervasive one and the concept has been extended to include governance as a fourth dimension (Herriman, Storey, Smith, & Collier, 2012). Governance is discussed in more detail in Chap. 5 of this book.

The concepts of well-being and Quality of Life (QoL) have always been central to discussions of sustainability with consensus around Costanza's (2009) view that we have forgotten that the economy is a means to achieve well-being rather than an end in itself (Jabareen, 2008; Parker, 2012; Scott, 2012). Definitions and discussions of sustainability increasingly highlight the importance of finding ways to improve human well-being globally as a central goal of sustainability (Bandarage, 2013). These discussions of sustainability and well-being are typically based on the idea that well-being is determined by access to, and use of, different forms of capital (Antrop, 2006; Costanza, 2009; Jabareen, 2008; Redclift, 2005; Scott, 2012). A well-being or capitals approach to sustainability argues that individuals, governments, businesses and other organisations need to consider the impacts of their decisions and impacts on all these capitals and that the maintenance and enhancement of all these different capitals is necessary (Costanza, 2009; Lehtonen, 2004). The main types of capital that are included in definitions and discussions of sustainability are.

- Natural capital, or the resources, amenities and assets available in the natural environment to support human life;
- Financial capital, defined as the monetary assets and resources available for investment and exchange;
- Built capital, or the physical infrastructure that supports production systems;
- Social capital, which refers to the value of trust and reciprocity that develops from networks and relationships;
- Human capital, or the skills, assets, knowledge, capabilities, and experiences of people;
- Political capital, which refers to the ability of individuals and groups to access and influence decisions that affect them; and
- Cultural capital, often defined as the traditions, knowledge, arts, rituals and languages that support values and identity (adapted from Moscardo, 2013a).

Criticisms common to both the TBL and capitals approach to sustainability are that they encourage people to think about the dimensions and forms of capital as separate and equal and to assume that finding a balance amongst them is desirable and/or easy (Antrop, 2006; Milne & Gray, 2013; Wikström, 2010). The assumption that the different forms of capital are equal in terms of value and substitutability, is referred to as a weak approach to sustainability (Dietz & Neumayer, 2007). This is contrasted with a strong approach to sustainability where natural capital is recognised as being unique and not substitutable and therefore sustainability strategies must specifically seek to maintain or increase natural capital (Hediger, 1999). Springett (2010) expands on this distinction arguing that weak approaches to sustainability not only treat all forms of capital as equal but also typically assume that growth is necessary, and that sustainability can be addressed through changes to current management processes, especially through a consideration of the TBL and concentration on eco-efficiency. Strong approaches to sustainability see natural capital as distinctive, replace the ideas of growth and eco-efficiency with considerations of justice, equity and enhancing capabilities, and argue that sustainability can only be addressed through major changes, especially to political systems (Springett, 2010).

Springett's (2010) expanded view of a strong approach to sustainability highlights the importance of ethics, justice and equity as key problems to be addressed. This reflects a broader shift in discussions of sustainability in which there is a more explicit consideration of the ethical foundations of decisions and a recognition that the current global economic system is not equitable (Redclift, 2005; Springett, 2010). Concepts such as justice, rights and responsibilities have been more closely examined in the context of sustainability (Parker, 2012) and Chap. 3 of this book, examines the ethical foundations of sustainability in more detail. Part of this shift from green to responsible has been the attention paid to the central role of CSR as a critical method for supporting sustainability (van Marrewijk, 2003). Chapter 4 of this book provides a more detailed analysis of CSR.

This increased attention to the concept of CSR is based on the idea of sustainability as both an integrative concept and a guiding ideal. For some authors (cf. Jabareen, 2008), sustainability can be conceptualised as either a new approach to strategy and management or a vision of utopia. The first perspective argues that sustainability requires integration across sectors and boundaries and the inclusion of a wider range of stakeholders than is usually considered in decisions (Jabareen, 2008). The second and more common perspective argues that the real value of sustainability is that it serves as a vision for a better future, a guiding ideal to drive change and a trigger for critical discussion about global issues (Jabareen, 2008; Parker, 2012; Redclift, 2005). Consistent with the shift towards ethics, sustainability has also been presented as a new moral ideal (Meijboom & Brom, 2012).

Central to this idea of sustainability as a guiding ideal is the final theme, the need for radical or transformative change to current practices in many areas (Dresner, 2008; Spring & Brauch, 2011). For many commentators this also suggests a need to move away from the use of the term sustainable development as the idea of development may be seen as inappropriate for such fundamental changes (Redclift, 2005).

Figure 2.1 provides a map setting out the major elements of sustainability and includes concepts often related to sustainability practice. A common starting point in discussions of changing business practice towards sustainability is to assess the current impacts an organisation or activity has on the various dimensions of sustainability and establish a monitoring system to measure and track sustainability performance (cf. Esty & Winston, 2009; Swallow, 2009). Tools such as ecological and carbon footprint analysis have become popular options for conducting sustainability audits and suggesting indicators for ongoing sustainability monitoring (Galli et al., 2012). Major programs for sustainability monitoring and reporting include the Global Reporting Initiative (GRI, 2014) and the United Nations Indicators of Sustainable Development (United Nations, 2007). Sustainability monitoring has become a major area of debate with considerable critical discussion about how and what is measured to determine sustainability performance (Davidson, 2011). Monitoring and reporting are, in turn, core elements of various environmental management systems (EMS) such as the ISO 14000 program (ISO, 2014) and the increasing number of eco-accreditation or sustainability labelling schemes, such as Fairtrade (Fairtrade Australia & New Zealand, 2014).

Another important aspect of sustainability measurement and assessment is the concept of life cycle analysis or assessment (LCA), also referred to as cradle to grave or cradle-to-cradle analysis. All these terms refer to "*thinking about the product from its very beginning, the resources and materials needed to create it, through its various production stages, its use by the customers and how it is disposed*" (Moscardo, 2013a, p. 21). Such an approach directs attention to the larger system in which production exists.

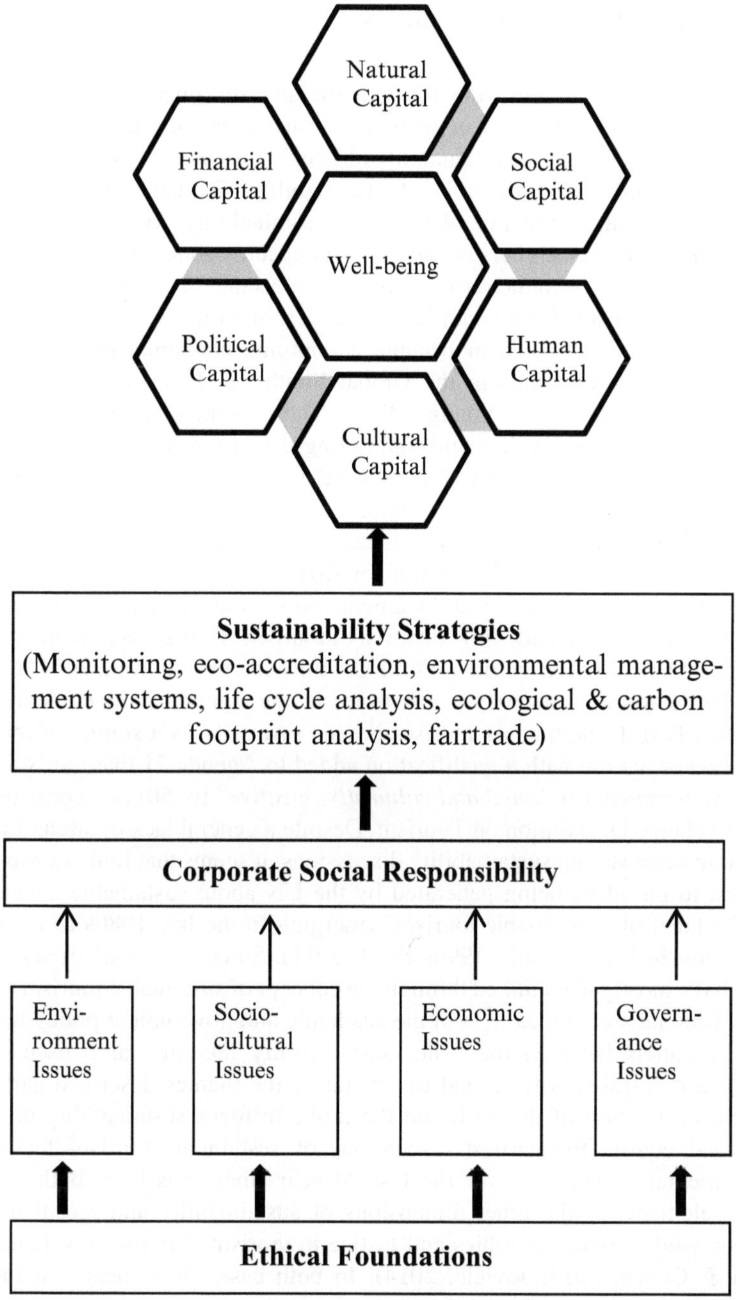

Fig. 2.1 A sustainability concept map (adapted from Moscardo, 2013b; Van Marrewijk, 2003)

2.4 Tourism and Sustainability

Table 2.2 provides a time line for tourism and sustainability matched to the previous one for sustainability in general. As can be seen there is a similar pattern of evolution with regard to sustainability although the tourism responses tend to lag behind the mainstream discussions of sustainability. The table highlights an early period prior to the emergence of the label sustainability, in which a number of commentators and researchers began to raise concerns about the negative environmental, social and economic costs of tourism, especially as the number of international tourists, mostly from countries in the Global North, rose dramatically after World War II and as tourism became a common economic development tool adopted by many countries in the Global South (cf. Krippendorf, 1982, 1987; Mathiesen & Wall, 1982; Young, 1973). These commentators questioned the value of tourism as a development tool noting that it rarely brought the promised economic benefits and often was associated with significant environmental and socio-cultural costs. A number of authors suggested the need for new ways of thinking about tourism introducing ideas such as "gentle" and "soft" tourism (Baumgartner, 1978; Scherle & Hopfinger, 2013).

The first UN conferences and documents on sustainability produced in the late 1980s and early 1990s made little mention of tourism. Tourism is only mentioned in passing in Our Common Future (World Commission on Environment & Development, 1987) and appears only a few times in the Agenda 21 document (United Nations, 1993). In both cases tourism is seen primarily as a source of income to support conservation with a qualification added to Agenda 21 that tourism needed to be "*environmentally sound and culturally sensitive*" (p. 50) and consistent with the 1989 Hague Declaration on Tourism. Despite a general lack of attention paid to tourism in main stream sustainability discussions, it seems that tourism was paying attention to the ideas being generated by the UN about sustainable development with the label of 'sustainable tourism' emerging in the late 1980s and becoming well established in the early 1990s (Hall, 2011). Since then tourism and sustainability have mostly been linked through the concept of sustainable tourism (ST) and ST has become a core idea in both the academic and government policy literature.

The parallels between the wider sustainability literature and discussions of tourism and sustainability extend to several of the themes described earlier. For example in the case of the TBL and the move to focus sustainability on justice, ethics and equity, the earliest discussions of sustainable tourism focussed on environmental issues and over the last 20 years there has been both a shift to include elements of the other dimensions of sustainability and recent increased attention paid to issues of ethics and justice in tourism (Bramwell & Lane, 2008; Cohen & Cohen, 2012; Jovicic, 2014). In both cases, it is clear that these are emerging rather than well-established themes in the tourism literature and are still not commonly considered in tourism practice (Jamal, Camargo, & Wilson, 2013; Spindler, 2013). Similarly, in recent years there has been greater attention paid to the connections between tourism and well-being or QoL and the various forms of

Table 2.2 A tourism and sustainability timeline

Phases	Sustainability	Tourism and sustainability
Rising awareness	1962 Carson's The Silent Spring 1965 First Fairtrade Organization 1968 Hardin writes The Tragedy of the Commons	1973 Tourism: Blessing or Blight? 1976 Doxey's Irridex 1982 Tourism, Economic, Physical and Social Impacts 1982 Towards new tourism policies 1987 The Holiday Makers
United Nations attention	1972 UN Conference in Stockholm sets up the UN Environment Programme (UNEP) 1987 Brundtland Report 1990 First report on climate change 1992 UN Earth Summit held in Rio de Janeiro 1993 Agenda 21 published 1993 First meeting of the UN Commission on Sustainable Development	1989 Hague Declaration on Tourism 1990 Action Strategy for Sustainable Tourism Development 1992 English Tourist Board produces The Green Light: A Guide to Sustainable Tourism 1993 First Issue of the Journal of Sustainable Tourism is published 1994 Green Globe launched 1995 Publication of Agenda 21 for the Travel & Tourism Industry 1995 World Conference on Sustainable Tourism 1999 WTO Global Code of Tourism Ethics
The New Millennium	2002 UN Millennium Summit in Johannesburg 2002 Global Reporting Initiative (GRI) releases first guidelines 2006 Stern report on Climate Change 2007 Al Gore and the IPCC win a Nobel Peace Prize	2001 First Building Excellence in Sustainable Tourism Education network (BEST EN) Think Tank is held in South Africa 2001 Establish rules for Biological Diversity & Sustainable Tourism 2002 UN International Year of Ecotourism 2002 the WTO launches Sustainable Tourism-Eliminating Poverty (STEP) 2003 Djerba Declaration on Tourism and Climate Change 2010 First Fairtrade tourism organization established in South Africa 2010 Norway launches national tourism strategy to become a carbon neutral destination 2013 GRI releases list of areas for indicators for tourism related activities

capital (Moscardo & Murphy, 2014). Arguably most attention is focussed on the well-being of tourists rather than of destination residents or tourism staff and discussions have tended to focus on the value of the various capitals for tourism, rather than the potential for tourism to make a positive contribution to the capitals of destinations (Moscardo, 2012). The central role of CSR in sustainability has also been discussed in the tourism literature but is still only rarely found in tourism business practice (Mihalic, 2014). Finally, there appears to be a similar concern in the tourism literature that sustainability requires fundamental and significant transformative change and a growing consensus that very little has changed in tourism practice (Bramwell and Lane, 2012, 2013; Buckley, 2012; Hall, 2011; Holden, 2009; Moscardo, 2011).

Tourism can also be examined in terms of its adoption of the sustainability strategies listed in Fig. 2.1. There is some evidence that all these strategies have been applied to tourism in some way, but arguably that adoption has either been very rare, as in the case of LCA (Castellani & Sala, 2012), very recent, as is the case with CSR (Sheldon & Park, 2011), or very fragmented, as in the case of eco-labels and eco-accreditation (Haaland & Aas, 2010). Sustainability monitoring is the strategy most widely discussed in tourism but this approach is subject to considerable criticism, as will be seen in the next sections.

2.5 Sustainable Tourism or Tourism for Sustainability

"*Despite the plethora of publications, conferences, and strategies that deal with sustainability, tourism is arguably less sustainable than it has ever been*" (Hall, 2010, p. 131). This is not an uncommon conclusion in academic discussions (Bramwell & Lane, 2013; Holden, 2009; Weaver, 2014) and there is a long history of critical concern about the way in which tourism researchers, policymakers and practitioners have conceptualised sustainability from early papers by Butler (1993), Wheeller (1993) and Wall (1997) and continuing to more recent papers by Hall (2009), Lane (2009), Buckley (2012), and Saarinen (2013). These critiques can be organised around four interconnected themes: ST as an end not a means, equating ST with alternative or eco-tourism, failures in tourism policy and planning, and challenges inherent in the nature of tourism itself.

2.5.1 Sustainable Tourism as an End Not a Means

Table 2.3 contains some of the most commonly used definitions of ST. Several features of these definitions have attracted substantial criticism. It has been suggested that these definitions reflect a weak approach to sustainability (Hunter, 2002; Liu, 2003) with assumptions about the desirability of growth, the pre-eminence of economic goals, and a focus on adjustments to current practices consistent with Springett's (2010) arguments. Saarinen (2013) refers to ST definitions as having a tourism first focus in which the needs of tourists and businesses are primary, the main goal is the maintenance of tourism itself and the suggested management changes are product or supply centred (Butler, 1999; Hall, 2011; Hardy, Beeton, & Pearson, 2002; McCool & Moisey, 2008; Saarinen, 2006; Sharpley, 2000). Implicit in this approach to ST is the assumption that some form of tourism will be sustainable rather than a willingness to acknowledge that no

Table 2.3 Commonly used definitions of sustainable tourism (ST)

Definition	Source
Sustainable tourism development meets the needs of present tourists and host regions while protecting and enhancing opportunities for the future.	WTO (1993), p. 7
Tourism that takes full account of its current and future economic, social and environmental impacts, addressing the needs of visitors, the industry, the environment and host communities	UNWTO (2015)
Tourism which is economically viable but does not destroy the resources on which the future of tourism will depend, most notably the physical environment and the social fabric of the host community	Swarbrooke (1999), p. 13
Tourism that respects both local people and the traveller, cultural heritage and the environment	UNESCO (2015)
Sustainable tourism is tourism committed to generating a low impact on the surrounding environment and community by acting responsibly while generating income and employment for the local economy and aiding social cohesion	Visit Scotland (2015)

tourism at all may be the only sustainable option (Moscardo, 2008b; Saarinen, 2013). This approach concentrates attention on the immediate, negative impacts that tourism can have on the destination with little consideration of the wider impacts of the global tourism system or the longer term changes associated with tourism to a destination (Butler, 1999; Hall, 2009; McCool & Moisey, 2008; Saarinen, 2006; Sharpley, 2000).

Also absent from these definitions of ST are considerations of resource use by tourism, awareness of the problems associated with using tourism as an economic development tool, acknowledgement of issues related to the North-South divide in tourism production and consumption, any consideration of demand, and a lack of awareness of the links between tourism and the QoL or well-being of destination communities (Butler, 1999; Hall, 2007; Holden, 2009; Liu, 2003; McCool & Moisey, 2008; Saarinen, 2006, 2013; Sharpley, 2000). The latter three issues have attracted particular critical attention.

According to Sharpley (2000, p. 11) patterns of international tourism "*reinforce rather than diminish global socio-economic inequities*" while Solomon (2005 cited in Hall, 2007, p. 114) argues that "*tourism is largely an avenue and instrument of the rich and affluent whose wealth has been accumulated in the context of unjust structures and systems of society*". These concerns about the failure of ST to address issues of justice and equity have contributed to calls for greater attention to be paid to the ethical nature of tourism as an activity (cf. Macbeth, 2005) and to the development of responsible tourism as an alternative to ST (cf. Jamal et al., 2013).

In turn this requires a more critical consideration of the nature of tourist demand and behaviours (Liu, 2003; Saarinen, 2006; Sharpley, 2000). Holden (2009) notes that while proponents of ST typically argue that tourists are seeking these new sustainable forms of tourism, there is very little evidence that this is the case. On the contrary, there is growing body of research to suggest that even those who are environmentally and socially responsible in their everyday lives see holidays as an opportunity to abandon those behaviours and values (Barr & Prillwitz, 2012; Miao & Wei, 2013), that few tourists have a clear idea of what responsible travel behaviours might be (Miller, Rathouse, Scarles, Holmes, & Tribe, 2010), and that generally tourists are reluctant to support restrictions to their travel behaviours (Gössling, Scott, Hall, Ceron, & Dubois, 2012). Not surprisingly, both tourism consumption and its negative global impacts continue to grow (Buckley, 2012).

Once the maintenance and growth of tourism becomes the primary focus of policy makers and practitioners, it seems easy to forget that for a destination the value in hosting tourism lies in its contributions to various aspects of community well-being or destination resident quality of life (McCool & Moisey, 2008; Moscardo, 2009; Saarinen, 2006, 2013). As previously noted, there has been an increase in attention paid to the links between tourism and QoL, but the majority of work in this area is still concerned with the well-being of tourists rather than destination communities or stakeholders. Thus, in the current conceptualisations of ST tourism has become the end goal rather than the means to an end.

2.5.2 Sustainable Tourism as Alternative or Ecotourism

Current conceptualisations of ST treat it as a form of tourism that has been modified to have fewer immediate destination specific negative impacts, rather than a fundamentally different approach to tourism. This reflects the historical development of the ST concept where it emerged from the dialectic between small-scale alternative forms of tourism and the idea of large-scale, mass, traditional tourism (Butler, 1999; Hardy et al., 2002; Liu, 2003; McCool & Moisey, 2008; Weaver, 2014). For a considerable time ST was equated with ecotourism and this continues in some areas despite the growing evidence of the failure of ecotourism to consistently contribute to the sustainability of destinations (Lane, 2009). This view of ST as an alternative to mass tourism was flawed in a number of ways but in terms of the present discussion of tourism sustainability the key problems are that:

- The bulk of tourism is not, and cannot be replaced by, alternative small-scale options so this approach fails to alter the sustainability of the large majority of tourist activities;
- Alternative tourism often relies upon the infrastructure and systems of traditional mass tourism and so contributes to the negative global impacts of tourism;
- The majority of cases demonstrating alternative tourism exist in peripheral regions and/or emerging destinations and are not appropriate for the addressing the problems of established destinations that already have mass tourism; and
- Alternative tourism ventures often act as precursors to more traditional development pathways for tourism (Butler, 1999; Hardy et al., 2002; Liu, 2003; McCool & Moisey, 2008; Weaver, 2014).

It has been suggested that it is possible to learn from these attempts to address tourism sustainability through the development of small-scale alternative tourism products. Weaver (2014), for example talks about an enlightened mass tourism, which has stronger ethical components, more extensive corporate social responsibility programs and greater attention paid to interpretation and tourist education. Liu (2003) offers a very different conclusion suggesting that because the evidence from evaluations of alternative tourism indicates that this approach merely spreads negative impacts further, the answer is to direct more tourists to places like Las Vegas, which are described as more impact-resilient.

2.5.3 Failures in Tourism Policy and Planning

Given that ST is seen as a variation of standard tourism practice, it is not surprising to find that there have been few, if any, changes in approaches to tourism policy and planning (Hall, 2011; McCool & Moisey, 2008). Moscardo's (2011) review of tourism planning models in textbooks and guides showed that there had been very little change in approaches to tourism planning since an earlier review by Getz in 1986. These reviews also highlight a number of features of tourism planning models that have been identified by others as major failures including:

- Limited opportunities for effective involvement by destination communities in tourism governance and practice (Byrd, 2007; Liu, 2003);
- That tourism planning is seen primarily as business planning focussed on marketing rather than as strategic community development (Ruhanen, 2010); and
- A focus on economic growth with little discussion of negative impacts (Hall, 2011).

There is also a lack of integration with other activities (Butler, 1999) reflecting what Hunter (2002, p. 4) calls "sectoral parochialism" in both tourism practice and research. Tourism planning and policy also fails in the area of sustainability monitoring (Butler, 1999). Partly this is because there is only a limited understanding of the processes that contribute to tourism impacts (McCool & Moisey, 2008), especially those in the social and cultural realm (Moscardo, Konovalov, Murphy, & McGehee, 2013). It also reflects a failure to more directly link tourism to its contributions to destination QoL (Saarinen, 2006) and an unwillingness by tourism policy makers to acknowledge the negative impacts of tourism (Hall, 2011).

2.5.4 Challenges Inherent to Tourism

The final theme in discussions of ST are those that seek to identify the characteristics of tourism that contribute to the challenges identified in the previous sections.

The most commonly cited characteristic is the fragmented and diverse nature of tourism, which crosses multiple government boundaries both within and across jurisdictions (Hall, 2011; Lane, 2009; McCool & Moisey, 2008). This makes it difficult to find sufficient authority to implement wide scale programs, challenging to involve all the relevant stakeholders in change processes and difficult to find change leaders. The parochialism of tourism is also an issue with Bramwell and Lane (2012) noting a strong tendency in tourism to adapt existing strategies rather than pursue innovation. Other issues include intense competition within and amongst tourism destinations (Frey & George, 2010), general denial of the negative impacts of tourism (Hall, 2011; Lane, 2009), weak links between academics and industry (Lane, 2009) and a widespread support for the idea that growth in tourism is a good thing (Lane, 2009). Moscardo (2009) describes a dominant or hegemonic social representation of tourism amongst both academics and the wider public in which being a tourist is seen as a worthwhile and highly desirable activity. Tourism is rarely considered as an "*ego-centric, status-driven activity, generated in part from the failure of individual tourists to find meaning in their daily lives, with only temporary benefits at best*" (Moscardo, 2009, pp 168–169), although some recent discussions have begun to suggest this as a realistic alternative view (McKercher, 2014).

2.6 Tourism and Sustainability: Conclusions

Despite the lack of real progress in improving the sustainability of tourism and in the development of tourism as a tool to support sustainability more generally, the continued growth in both tourism and its contribution to climate change and other major sustainability problems creates an imperative to keep trying to change tourism towards sustainability (Bramwell & Lane, 2008; Saarinen, 2013). While some have suggested abandoning the label of sustainable tourism (Butler, 1999; Moscardo, 2008a, 2008b), it is might be reasonable to retain it but clarify its use "as shorthand for tourism's contribution to sustainable development" (Hunter, 2002, p. 12). Suggestions for improving tourism's contributions to sustainable development include:

1. The need to look outside tourism for new ideas (Bramwell & Lane, 2012);
2. More explicit acknowledgement of the social justice and equity issues related to tourism production and consumption (Bramwell & Lane, 2008);
3. A focus on limiting growth which means more research into the nature of demand and tourist behaviour (Bramwell & Lane, 2013; Hall, 2009; Saarinen, 2013);
4. Greater attention to integration with other activities and a shift in focus away from the QoL of tourists towards the QoL of destination residents (Bramwell & Lane, 2012; Moscardo & Murphy, 2014; Saarinen, 2006); and
5. More attention to the concept of risks associated with tourism development (Bramwell & Lane, 2008).

There is a complex relationship between tourism and sustainability and a widespread consensus that the key to improving this relationship is the development of a greater sense of responsibility towards and within the various stakeholders in the tourism system. According to Jamal (2004) this development of 'good' tourism requires a change in the way we think and teach about tourism. The review of tourism and sustainability in this chapter highlights the need to improve education for sustainability in tourism.

References

Antrop, M. (2006). Sustainable landscapes: Contradiction, fiction or utopia. *Landscape & Urban Planning, 75*, 187–197.

Bandarage, A. (2013). *Sustainability and well-being*. Basingstoke: Palgrave Macmillan.

Barkemeyer, R., Figge, F., Holt, D., & Wettstein, B. (2009). What the papers say: Trends in sustainability. A comparative analysis of 115 leading national newspapers worldwide. *Journal of Corporate Citizenship, 2009*(33), 68–86.

Barr, S., & Prillwitz, J. (2012). Green travellers? Exploring the spatial context of sustainable mobility styles. *Applied Geography, 32*, 798–809.

Baumgartner, F. (1978). Le tourisme dans le Tiers monde—Contribution au developpement? *Tourist Review, 3*(1), 14–18.

Bramwell, B., & Lane, B. (2008). Priorities in sustainable tourism research. *Journal of Sustainable Tourism, 16*(1), 1–4.

Bramwell, B., & Lane, B. (2012). Towards innovation in sustainable tourism research? *Journal of Sustainable Tourism, 20*(1), 1–7.

Bramwell, B., & Lane, B. (2013). Getting from here to there: Systems change, behavioural change and sustainable tourism. *Journal of Sustainable Tourism, 21*(1), 1–4.

Brundtland, G. H. (1987). *Our common future*. Oxford: Oxford University Press.

Buckley, R. (2012). Sustainable tourism: Research and reality. *Annals of Tourism Research, 39*(2), 528–546.

Butler, R. W. (1993). Tourism—An evolutionary perspective. In J. G. Nelson, R. W. Butler, & G. Wall (Eds.), *Tourism and sustainable development: Monitoring, planning, managing*. Waterloo: University of Waterloo, Department of Geography.

Butler, R. W. (1999). Sustainable tourism: A state-of-the-art review. *Tourism Geographies, 1*(1), 7–25.

Byrd, E. T. (2007). Stakeholders in sustainable tourism development and their roles: Applying stakeholder theory to sustainable tourism development. *Tourism Review, 62*(2), 6–13.

Castellani, V., & Sala, S. (2012). Ecological footprint and life cycle assessment in the sustainability assessment of tourism activities. *Ecological Indicators, 16*, 135–147.

Cohen, E., & Cohen, S. A. (2012). Current sociological theories and issues in tourism. *Annals of Tourism Research, 39*(4), 2177–2202.

Costanza, R. (2009). Toward a new sustainable economy. *Real-World Economics Review, 49*, 20–21.

Davidson, K. M. (2011). Reporting systems for sustainability: What are they measuring? *Social Indicators Research, 100*(2), 351–365.

Desjardins, J. R. (2007). *Business, ethics and the environment*. Upper Saddle River, NJ: Pearson.

Dietz, S., & Neumayer, E. (2007). Weak and strong sustainability in the SEEA: Concepts and measurement. *Ecological Economics, 61*(4), 617–626.

Dresner, S. (2008). *The principles of sustainability*. London: Earthscan.

Elkington, J. (2004). Enter the Triple Bottom Line. In A. Henriques & J. Richardson (Eds.), *The triple bottom line: Does it all add up? Assessing the sustainability of business and CSR* (pp. 1–16). London: Earthscan.

Esty, D., & Winston, A. (2009). *Green to gold.* Hoboken: Yale University Press.

Fairtrade Australia and New Zealand. (2014). *What is fairtrade?* Available online: http://fairtrade.com.au/page/what-fairtrade. Accessed 29 Sept 2014.

Frey, N., & George, R. (2010). Responsible tourism management: The missing link between business owners' attitudes and behaviour in the Cape Town tourism industry. *Tourism Management, 31*(5), 621–628.

Galli, A., Wiedmann, T., Ercin, E., Knoblauch, D., Ewing, B., & Giljum, S. (2012). Integrating ecological, carbon and water footprint into a "footprint family" of indicators: Definition and role in tracking human pressure on the planet. *Ecological Indicators, 16*, 100–112.

Getz, D. (1986). Models in tourism planning: Towards integration of theory and practice. *Tourism Management, 41*(3), 287–295.

Global Reporting Initiative. (2014). *About GRI.* Available online: https://www.globalreporting.org/information/about-gri/Pages/default.aspx. Accessed 15 Oct 2014.

Gössling, S., Scott, D., Hall, C. M., Ceron, J. P., & Dubois, G. (2012). Consumer behaviour and demand response of tourists to climate change. *Annals of Tourism Research, 39*(1), 36–58.

Haaland, H., & Aas, Ø. (2010). Eco-tourism certification—Does it make a difference? A comparison of systems from Australia, Costa Rica and Sweden. *Scandinavian Journal of Hospitality and Tourism, 10*(3), 375–385.

Hall, C. M. (2007). Pro-poor tourism: Do 'tourism exchanges benefit primarily the countries of the South'? *Current Issues in Tourism, 10*(2–3), 111–118.

Hall, C. M. (2009). Degrowing tourism: Décroissance, sustainable consumption and steady-state tourism. *Anatolia, 20*(1), 46–61.

Hall, C. M. (2010). Changing paradigms and global change: From sustainable to steady-state tourism. *Tourism Recreation Research, 35*(2), 131–143.

Hall, C. M. (2011). Policy learning and policy failure in sustainable tourism governance: From first-and second-order to third-order change? *Journal of Sustainable Tourism, 19*(4–5), 649–671.

Hardy, A., Beeton, R. J., & Pearson, L. (2002). Sustainable tourism: An overview of the concept and its position in relation to conceptualisations of tourism. *Journal of Sustainable Tourism, 10* (6), 475–496.

Hediger, W. (1999). Reconciling "weak" and "strong" sustainability. *International Journal of Social Economics, 32*(3), 481–492.

Herriman, J., Storey, H., Smith, P., & Collier, G. (2012). Working relationships for sustainability. *Commonwealth Journal of Local Governance*, 10. Available online http://epress.lib.uts.edu.au/ojs/index.php/cjlg. Accessed 17 Mar 2014.

Holden, A. (2009). The environment-tourism nexus: Influence of market ethics. *Annals of Tourism Research, 36*(3), 373–389.

Hunter, C. (2002). Aspects of the sustainable tourism debate from a natural resources perspective. In R. Harris, T. Griffin, & P. Williams (Eds.), *Sustainable tourism* (pp. 3–23). Oxford: Elsevier Science.

International Standards Organization. (2014). ISO 14000—Environmental management. Available online: http://www.iso.org/iso/home/standards/management-standards/iso14000.htm. Accessed 29 Sept 2014.

Jabareen, Y. (2008). A new conceptual framework for sustainable development. *Environment, Development and Sustainability, 10*(2), 179–192.

Jamal, T. B. (2004). Virtue ethics and sustainable tourism pedagogy: Phronesis, principles and practice. *Journal of sustainable tourism, 12*(6), 530–545.

Jamal, T., Camargo, B. A., & Wilson, E. (2013). Critical omissions and new directions for sustainable tourism: A situated macro–micro approach. *Sustainability, 5*(11), 4594–4613.

Jovicic, D. Z. (2014). Key issues in the implementation of sustainable tourism. *Current Issues in Tourism, 17*(4), 297–302.

Klein-Vielhauer, S. (2009). Framework model to assess leisure and tourism sustainability. *Journal of Cleaner Production, 17*, 447–454.

Kolk, A. (2008). Sustainability, accountability and corporate governance: Exploring multinationals' reporting practices. *Business Strategy and the Environment, 17*(1), 1–15.

Krippendorf, J. (1982). Towards new tourism policies: The importance of environmental and socio-cultural factors. *Tourism Management, 3*, 135–148.

Krippendorf, J. (1987). *The holiday makers: Understanding the impact of leisure and travel.* Oxford: Butterworth-Heinemann.

Lamberton, G. (2005). Sustainability accounting—A grief history and conceptual framework. *Accounting Forum, 29*(1), 7–26.

Lane, B. (2009). Thirty years of sustainable tourism: Drivers, progress, problems—And the future. In S. Gossling, C. M. Hall, & D. B. Weaver (Eds.), *Sustainable tourism futures* (pp. 19–32). New York: Routledge.

Lehtonen, M. (2004). The environmental-social interface of sustainable development: Capabilities, social capital, institutions. *Ecological Economics, 49*, 199–214.

Liu, Z. (2003). Sustainable tourism development: A critique. *Journal of Sustainable Tourism, 11*(6), 459–475.

Lozano, R. (2008). Envisioning sustainability three-dimensionally. *Journal of Cleaner Production, 16*, 1838–1846.

Lumley, S., & Armstrong, P. (2004). Some of the nineteenth century origins of the sustainability concept. *Environment, Development & Sustainability, 6*(3), 367–378.

Macbeth, J. (2005). Towards an ethics platform for tourism. *Annals of Tourism Research, 32*(4), 962–984.

Mathiesen, A., & Wall, G. (1982). *Tourism: Economic, physical and social impacts.* London: Longman.

McCool, S. F., & Moisey, R. N. (2008). Introduction: Pathways and pitfalls in the search for sustainable tourism. In S. F. McCool & R. N. Moisey (Eds.), *Tourism, recreation and sustainability* (2nd ed., pp. 1–16). Wallingford: CABI.

McKercher, B. (2014). Tourism: The quest for the selfish. *Tourism Recreation Research, 39*(1), 99–104.

Mebratu, D. (1998). Sustainability and sustainable development: Historical and conceptual review. *Environmental Impact Assessment Review, 18*(6), 493–520.

Meijboom, F. L., & Brom, F. W. (2012). Ethics and sustainability: Guest or guide? On sustainability as a moral ideal. *Journal of Agricultural and Environmental Ethics, 25*(2), 117–121.

Miao, L., & Wei, W. (2013). Consumers' pro-environmental behaviour and the underlying motivations: A comparison between household and hotel settings. *International Journal of Hospitality Management, 32*, 102–112.

Mihalic, T. (2014). Sustainable-responsible tourism discourse—Towards responsustable tourism. *Journal of Cleaner Production*. Available online: http://www.sciencedirect.com/science/article/pii/S0959652614013596. Accessed 8 Jan 2015.

Miller, G., Rathouse, K., Scarles, C., Holmes, K., & Tribe, J. (2010). Public understanding of sustainable tourism. *Annals of Tourism Research, 37*(3), 627–645.

Milne, M. J., & Gray, R. (2013). W(h)ither ecology? The triple bottom line, the global reporting initiative, and corporate sustainability reporting. *Journal of Business Ethics, 118*(1), 13–29.

Moscardo, G. (2008a). Introduction. In G. Moscardo (Ed.), *Building community capacity for tourism* (pp. ix–xii). Wallingford: CABI.

Moscardo, G. (2008b). Sustainable tourism innovation: Challenging basic assumptions. *Tourism and Hospitality Research, 8*(1), 4–13.

Moscardo, G. (2009). Tourism and quality of life: Towards a more critical approach. *Tourism and Hospitality Research, 9*(2), 159–170.

Moscardo, G. (2011). Exploring social representations of tourism planning: Issues for governance. *Journal of Sustainable Tourism, 19*(4–5), 423–436.
Moscardo, G. (2012). Building social capital to enhance the quality-of-life of destination residents. In M. Uysal, R. Perdue, & M. J. Sirgy (Eds.), *Handbook of tourism and quality-of-life research* (pp. 403–421). Dordrecht: Springer.
Moscardo, G. (2013a). Sustainability, economy and society. In G. Moscardo, G. Lamberton, G. Wells, W. Fallon, P. Lawn, A. Rowe, & W. Kershaw (Eds.), *Sustainability in Australian business: Principles and practice* (pp. 1–34). Brisbane: Wiley-Blackwell.
Moscardo, G. (2013b). Business ethics, corporate governance and corporate social responsibility. In G. Moscardo, G. Lamberton, G. Wells, W. Fallon, P. Lawn, A. Rowe, & W. Kershaw (Eds.), *Sustainability in Australian business: Principles and practice* (pp. 67–97). Brisbane: Wiley-Blackwell.
Moscardo, G., Konovalov, E., Murphy, L., & McGehee, N. (2013). Mobilities, community well-being and sustainable tourism. *Journal of Sustainable Tourism, 21*(4), 532–556.
Moscardo, G., & Murphy, L. (2014). There is no such thing as sustainable tourism: Re-conceptualizing tourism as a tool for sustainability. *Sustainability, 6*(5), 2538–2561.
Parker, K. A. (2012). Introduction—Sustainability in higher education. In K. A. Bartels & K. A. Parker (Eds.), *Teaching sustainability teaching sustainably* (pp. 1–16). Sterling: Stylus.
Paul, B. D. (2008). A history of the concept of sustainable development: Literature review. *Annals of the University of Oradea, Economic Sciences Series, 17*(2), 576–580.
Redclift, M. (2005). Sustainable development (1987–2005): An oxymoron comes of age. *Sustainable Development, 13*(4), 212–227.
Ruhanen, L. (2010). Where's the strategy in tourism strategic planning? Implications for sustainable tourism destination planning. *Journal of Travel and Tourism Research, 10*(1/2), 58–76.
Saarinen, J. (2006). Traditions of sustainability in tourism studies. *Annals of Tourism Research, 33* (4), 1121–1140.
Saarinen, J. (2013). Critical sustainability: Setting the limits to growth and responsibility in tourism. *Sustainability, 6*(1), 1–17.
Scherle, N., & Hopfinger, H. (2013). German perspectives on tourism geography. In J. Wilson & S. A. Clave (Eds.), *Geographies of tourism: European research perspectives* (pp. 69–91). Bingley: Emerald.
Scott, K. (2012). *Measuring wellbeing: Towards sustainability*. Abingdon: Routledge.
Scott, D., Hall, C. M., & Gossling, S. (2012). *Tourism and climate change: Impacts, adaptation and mitigation*. Abingdon: Routledge.
Sharpley, R. (2000). Tourism and sustainable development: Exploring the theoretical divide. *Journal of Sustainable Tourism, 8*(1), 1–19.
Sheldon, P. J., & Park, S. Y. (2011). An exploratory study of corporate social responsibility in the US travel industry. *Journal of Travel Research, 50*(4), 392–407.
Spindler, E. A. (2013). The history of sustainability. In I. Jenkins & R. Schroder (Eds.), *Sustainability in tourism: A multidisciplinary approach* (pp. 9–32). Berlin: Springer.
Spring, Ú. O., & Brauch, H. G. (2011). Coping with global environmental change—Sustainability revolution and sustainable peace. In H. G. Brauch, U. O. Spring, C. Mesjasz, J. Grin, P. -Kameri-Mbote, B. Chourou, P. Dinay, & J. Brokmann (Eds.), *Coping with global environmental change, disasters and security* (pp. 1487–1503). Berlin: Springer.
Springett, D. (2010). Education for sustainability in the business studies curriculum: Ideological struggle. In P. Jones, D. Selby, & S. Sterling (Eds.), *Sustainability education: Perspectives and practice across higher education* (pp. 75–92). Abingdon: Earthscan.
Swarbrooke, J. (1999). *Sustainable tourism management*. Wallingford: CABI.
Swallow, L. (2009). *Green business practices for dummies*. Hoboken: Wiley.
UNESCO. (2015). *Sustainable tourism*. Available at http://www.unesco.org/education/tlsf/mods/theme_c/mod16.html. Accessed 4 Feb 2015.
United Nations. (1993). *United Nations sustainable development*. Available online: http://sustainabledevelopment.un.org/content/documents/Agenda21.pdf. Accessed 21 Sept 2013.

United Nations. (2007). *Indicators of sustainable development: Guidelines and methodologies* (3rd ed.). New York: United Nations.
UNWTO. (2015). *Sustainable development of tourism*. Available at http://sdt.unwto.org/content/about-us-5. Accessed 4 Feb 2015.
Van Marrewijk, M. (2003). Concepts and definitions of CSR and corporate sustainability: Between agency and communion. *Journal of Business Ethics, 44*(2–3), 95–105.
Visit Scotland. (2015). *What is sustainable tourism?* Available online http://www.visitscotland.org/business_support/sustainable_tourism/what_is_sustainable_tourism.aspx. Accessed 8 Jan 2015.
Wall, G. (1997). Sustainable tourism—Unsustainable development. In S. Wahab & J. J. Pigram (Eds.), *Tourism development and growth* (pp. 33–49). London: Routledge.
Weaver, D. B. (2014). Asymmetrical dialectics of sustainable tourism: Toward enlightened mass tourism. *Journal of Travel Research, 53*(2), 131–140.
Wells, G. (2013). Introduction. In G. Wells (Ed.), *Sustainable business* (pp. 1–18). Cheltenham: Edward Elgar.
Wheeller, B. (1993). Sustaining the ego. *Journal of Sustainable Tourism, 1*(2), 121–129.
Wikström, P. A. (2010). Sustainability and organizational activities—Three approaches. *Sustainable Development, 18*(2), 99–107.
World Commission on Environment and Development. (1987). *Our common future*. Oxford: Oxford University Press.
WTO. (1993). *Sustainable tourism development: Guide for local planners*. Madrid: WTO.
Young, G. (1973). *Tourism: Blessing or blight?* Harmondsworth: Penguin.

United Nations. (2007). *Indicators of sustainable development: Guidelines and methodologies* [illegible] New York: United Nations.
UNWTO. (2005). *Sustainable development of tourism* [illegible]
[illegible]

Chapter 3
Ethics in Tourism

David A. Fennell

Abstract This chapter argues that ethics provides an alternative way in which to better understand an act upon tourism industry dilemmas (as an alternative to the over-reliance on the impacts literature). There is a rich foundation of ethics knowledge emerging in the tourism literature, and this foundation continues to expand at a modest rate. In an effort to present this knowledge in an organised manner, six main sections have been developed: (1) foundations and theories of ethics and tourism; (2) applied ethics and tourism; (3) types of tourism and ethics; (4) ethics in tourism education and research; (5) sustainability ethics; and (6) tourism and animal ethics. The chapter concludes by suggesting that there is often a disconnect between theory and practice when it comes to ethics. This disconnect makes it especially important for educators to adopt moral theory in the curriculum in developing the proper character for learners who may later become practitioners.

Keywords Applied ethics • Theoretical ethics • Impacts • Values • Animal ethics • Sustainability ethics

3.1 Introduction

Through four decades of tourism inquiry, the focus of scholarship has weighed heavily on the social, economic, and environmental impacts of tourism on and between people, destinations, and the natural world. This emphasis on impacts is visible in many if not all of our most important and time-tested works in the tourism literature (e.g., Butler, 1980; Doxey, 1975). However, when it comes to understanding why and how these problems take place, it is debatable whether such an overwhelming focus on impacts is the only, or even best, way to address these issues (McKercher, 1993).

Increasingly, tourism scholars are starting to view theoretical and applied ethics as a more proactive way of tackling tourism industry problems. If ethics has been defined as what is good or bad, or right or wrong in/for people, tourism ethics can be

---

D.A. Fennell (✉)
Department of Tourism Management, Brock University, St. Catharines L2S 3A1, Ontario, Canada
e-mail: dfennell@brocku.ca

G. Moscardo, P. Benckendorff (eds.), *Education for Sustainability in Tourism*, CSR, Sustainability, Ethics & Governance, DOI 10.1007/978-3-662-47470-9_3

defined as what is good or bad, right or wrong in/for tourism (Fennell, 2006a). This means that all of the issues discussed from the context of the impacts platform in tourism over the years (above) are also, and perhaps more importantly, moral issues. For example, recent attempts to build sustainability and responsibility into tourism are certain to be more successful if they are grounded in moral theory. We might yield more meaning (and application) in responsible tourism if scholars were prepared to source the over 2500 years of discourse on ethics in the way other applied fields like medicine, environmental, and marketing and business have. Ignoring this literature serves only to limit our ability to more effectively navigate through the labyrinth of tourism's many human-environment issues.

Ethics was not a topic of interest amongst tourism scholars until the early 1990s. New studies emerged at this time as a result of the AIEST congress in Paris in 1992, as well as the Rio Earth Summit also in 1992. These meetings catalysed a number of publications that emerged from 1993 onwards, including, for example, Lea (1993), D'Amore (1993), Wight (1993), Ahmed, Krohn, and Heller (1994); and Fennell and Malloy (1995). Major works that galvanised tourism ethics in the tourism field include Smith and Duffy (2003), Fennell (2006a), and Fennell and Malloy (2007), with at least three new monographs by tourism scholars forthcoming. These works have been complemented with an increasing number of academic articles in a range of different tourism journals.

Tightly connected to ethics are values. Values are essentially an outward manifestation of what we believe to be important. Economic prosperity as a primary goal for individuals and organisations, pleasure at all costs, the exploitation of children as acceptable if it saves money, and bribery and nepotism, are expressions of values held by individuals and groups within society. Part of the problem in the relationship between tourism industry practices and tourist participation is the lack of knowledge on the part of the latter about how tourism products are developed, and who stands to gain and at what cost. As observed by D'Sa (1999):

> There is a fundamental schism in tourism between market values (profit, competition, survival of the fittest) and community values (cooperation, care for the weakest, spirituality, and so on). This backwards prioritisation of values manifests itself in the failure of tourists to take responsibility for their actions.

It would seem fair to suggest, although more research is needed in this area, that most tourists enlist either preference or consensus values when they travel. This means that tourists value something (like a mass tourism experience) because they simply like it, with little cognitive thought to the inner workings of the tourism industry. They may also value travel (consensus values) because others may have an influence on their travel decisions and behaviours. These are different than consequence values (i.e., where decisions like travel are based on rational thought in consideration of outcomes), or authentic values where value is not based on what one likes, what others like, or what science tells me, but rather on will, authenticity, and faith (see Hodgkinson, 1983).

3.2 Key Reference Points for Tourism and Ethics

The expanding base of literature on ethics and tourism can be partitioned into six rather unique areas of emphasis. These include (1) foundations and theories of ethics and tourism; (2) applied ethics and tourism; (3) types of tourism and ethics; (4) ethics in tourism education and research; (5) sustainability ethics; and (6) tourism and animal ethics. The fifth section on sustainability ethics is not, at present, a topic that has been discussed by tourism scholars. I use it here because it has the potential to add to the continuing discourse on tourism, ethics and sustainability (see the accompanying powerpoint presentation on this topic), and because it has particular relevance to the subject of the present book. After a brief introduction to each of these six sections, many representative works are emphasised for the purpose of casting light on how this body of knowledge has relevance to tourism studies, ethics and education.

3.2.1 Foundations/Theories of Ethics and Tourism

Good representations of the foundations and theories of ethics can be found in a number of journals, and more recently books, which stress the importance of ethics in tourism. These works discuss many of the main theories of ethics, while some papers discuss criticisms of focussing on tourism ethics in tourism. Macbeth (2005) and Fennell (2006a) both argue that ethics must play a more important part in the development of tourism, both in the classroom and in the field. This, they argue, represents the maturation of tourism scholarship and practice. Caton (2012) carries many of these ideas forward on why morality should play a more important role in tourism, with the conclusion that light-hearted pleasures often lead to dire social consequences.

The breadth of tourism research has expanded to cover the broad theoretical terrain of ethics. This includes work on normative theories of ethics including deontological ethics and teleological ethics (including virtue ethics). Deontological theories (i.e., what is moral is that which abides by rules and principles, rather then what is seen as the proper ends), include Kantian ethics, social contract ethics, and religious sources such as the bible. Examples in the literature include Przeclawski (1996) on the deontology of tourism; Heintzman (1995) on the Golden Rule, and Lovelock (2008) on tourism and human rights.

Teleological theories focus on the right ends of human action. An act is seen to be good or bad, therefore, depending upon the consequences of its performance—there is a focus on achieving the optimum outcomes. Two forms of teleological theory include utilitarianism (maximising pleasure and minimising pain for the group) and hedonism (maximising pleasure and minimising pain for the individual). Dobson (2011) employed utilitarian theory in arguing that the tourism industry needs to be more effective at balancing the costs and benefits of tourism for all (both humans and animals). Fennell (2009) argues that it is indeed unfortunate that concepts such as pleasure and hedonism, despite how important they are in tourism, are manifestly underrepresented in the literature. Justice has recently become a topic of great interest in tourism. Examples include the work of Higgins-Desbiolles (2008) on globalisation and justice, as well as Hultsman (1995) on just tourism. Lee and Jamal (2008) argue that tourism tends to exacerbate many of the environmental justice issues such as lack of water or food, which plague many of the world's marginal communities. Instead of fixing these disparities, tourism tends to make them worse. Virtue ethics theory is also being explored in tourism studies. Baptista (2012) investigates the dimensions of a virtuous tourist in the context of a Mozambican village (see also Tribe, 2002, below).

Existentialism, as a subjectivist form of ethics, is rather less well emphasised in tourism research. Fennell (2008) used Kierkegaard's ethic of care to cast more light on the meaning of responsibility in tourism, and Brown (2013) uses Heideggerian phenomenology and Sartrean existentialism to examine how tourism might play a role in revealing individual authenticity—that tourism might play a role in moving away from the inauthenticity of everyday life.

It is also worth noting that some theorists take issue with the proposition that ethics is a better way forward for tourism. Gibson (2010) argues that we should not focus on the binaries inherent in an ethical focus, but rather on embodiment, emotions and sensory encounters in dissecting power relations and care (see also Butcher, 2009 who argues against ethics in tourism).

3.2.2 Applied Ethics and Tourism

A second main area of emphasis on tourism and ethics focuses on the applied. Codes of ethics, corporate social responsibility, and environmental ethics are examples of this work. A number of articles have been written on the value of codes of ethic for tourism. Payne and Dimanche (1996) argue for the use of codes of ethics in tourism because (1) the tourism industry must recognize that its basis is a limited resource, and that sustainable economic development requires limits to growth; (2) the tourism industry must realize that it is community-based, and that greater consideration must be given to the socio-cultural costs of tourism development; and (3) the tourism industry must also recognize that it is service-oriented, and that it must treat employees and customers ethically (p. 997). General overviews can be found in D'Amore (1993) and Mason and Mowforth (2006). Malloy and Fennell (1998) used content analysis methodology to deconstruct 414 separate code of ethics guidelines according to deontological and teleological messages. The same methodology was used in the context of the global whale watching industry (see Garrod & Fennell, 2004). Stonehouse (1997) provides an excellent overview of the use of codes of ethics in the Arctic.

Studies on corporate social responsibility have also become more numerous in the literature. The focus in these studies is on how corporations can implement policies and procedures that make them more socially and environmentally responsible—responsible actors who are behaving in a more sustainable fashion. Studies can be found in many sectors, including aviation (Ravinder, 2007), hotels (Holcomb, Upchurch, & Okumus, 2007), and the destination more generally (Williams, Gill, & Ponsford, 2007).

Tourism scholars have also been active in investigating aspects of ethics as they apply to the business and marketing of tourism enterprises. Examples include ethical work climate in lodging (Upchurch & Ruhland, 1995); ethical issues confronting travel agents (Dunfree & Black, 1996); and the management of ethics in the tourism supply chain (Keating, 2009). Weeden (2001) discusses at length the competitive advantage that tourism practitioners may realise by adopting ethical strategies. In the context of marketing, Yaman and Gurel (2006) focus on varying perspectives that tourism marketers embrace in delivering their products, and Hudson and Miller (2005) discuss the mechanics of responsible marketing from the perspective of Canadian Mountain Holidays. Holden (2003, 2009) is the chief proponent for the development of a new environmental ethics for tourism. His work has been influential in providing a basis from which to measure the negative and positive aspects of tourism development. For example, he found that even though the Cairngorms were susceptible to significant ecological impacts from downhill ski developments, human economic priorities are far more important than the fragile and unique nature of this region (Holden, 1999).

3.2.3 Types of Tourism and Ethics

Ethics has also played an important role in examining the characteristics of different forms of tourism, and how these types may differ from one another. Ecotourism is well represented in this research because of the oft-quoted contention that it exists as the most ethical form of tourism, i.e., it stimulates local participation and benefits, enhances conservation, and educates tourists.

Wight (1993) examined the marketing aspects of ecotourism as eco-ethics or eco-sell, and found motivations of industry stakeholders embody the eco-sell platform much more than eco-ethics. Karwacki and Boyd discussed ecotourism on primarily utilitarian grounds (egocentrism and issues of justice are touched on as well). Opponents of ecotourism argue that ecotourism can be criticised using utilitarianism calculus because benefits (e.g., economic impact) do not outweigh the costs. Pollution, loss of culture, displacement of local people, and so on, are disturbances that simply cannot be offset by marginal benefits realised by many on the periphery of the industry. By contrast, proponents argued that if ecotourism is well managed it has many positive spinoffs that are good for the environment and local people that cannot be realised through other mainstream forms of tourism like mass tourism. Fennell and Malloy (1995) built upon this work in the development of a comprehensive ethical framework that all ecotourism stakeholders could use in making good (teleology), right (deontology) and authentic (existentialism) decisions.

The ethics of third world tourism continue to be a topic of rich debate in tourism (see for example Ahmed et al., 1994; Hudson, 2007; Lea, 1993), and this has direct relevance to ethics and social tourism. In this latter capacity, Higgins-Desbiolles (2006) has written on tourism as a powerful social force, while Minnaert, Maitland, and Graham (2006) provide insight on the foundations of social tourism.

Various forms of responsible tourism have pushed the discussion on ethics and tourism laterally, but this research, in general, is hampered because theorists have not consistently used moral theory in their discussions. Pro-poor tourism (Harrison, 2008; Scheyvens, 2007) and fair trade tourism (Boluk, 2011; Nicholls, 2010) are noteworthy examples of this area of research. More specific examples of ethics and tourism types are found in the literature, and include poverty and slum tourism (Dyson, 2012; Selinger, 2009); the ethics of backpacker tourism (Speed, 2008); sex tourism (Eades, 2009; Wonders & Michalowski, 2001); and ethics and medical tourism (Meghani, 2011).

3.2.4 Ethics in Tourism Education and Research

Almost a decade ago, Fennell (2006a) commented that if tourism students get any form of instruction on ethics, such would almost certainly be as an elective or part of a core and context class in humanities for the purpose of fulfilling degree requirements. Tourism students in business faculties may be required to take a business ethics course as part of the curriculum (see Yaman, 2003 on business ethics and tourism in the context of naturalistic ethics).

Research on tourism education has progressed steadily in recent years from some of the earliest papers written on the topic at the onset of the 1990s (see for example Enghagen, 1990; O'Halloran, 1991). By the turn of the century, scholars began to place even more emphasis on the value of ethics in hospitality (Yeung, 2004) and tourism education. Tribe (2002) and later Jamal (2004), both suggest that we must more formally recognize the intrinsic nature of good in the Aristotlean fashion. That is, being virtuous (traits of character that make one a good person), if cultivated through good habits, allow us not only to do things correctly, but, most importantly, to do them for the right reasons. This may be accomplished through Aristotle's notion of phronesis (the development and exercise of practical wisdom) which allows for the hierarchal practice of more responsible forms and approaches to tourism.

The program that has perhaps had the biggest impact on tourism and education front is TEFI—the Tourism Education Futures Initiative. The aim of TEFI is 'to provide vision, knowledge, and a framework for tourism education programs that promote global citizenship and optimism for a better world' (Sheldon, Fesenmaier, & Tribe, 2011: 2). At the heart of the TEFI initiative are a set of fundamental values for tourism education, including stewardship, ethics, knowledge, mutuality and professionalism. Scholars who have used the TEFI template to move the tourism education agenda forward include Moscardo and Murphy (2011), Gretzel, Isacsson, Matarrita, and Wainio (2011), and Barber (2011). Other scholars have debated the need for a universal curriculum in tourism studies. Baum (2001) argues that such is still evolving, and is being heavily influenced by globalisation. Tourism scholars have also examined whether or not a code of ethics is required for research in the tourism field, and what it would look like (Moscardo, 2010). Other theorists have questioned the value of the audit culture that many of the contemporary western universities are adopting, and the impact this is having on the creation and organisation of knowledge in our field (Fennell, 2013).

3.2.5 Sustainability Ethics

Bridging from the previous section on ethics in tourism education, sustainability ethics calls for the need to incorporate ethics as a central driving force in all of our discussions and practices on sustainability. Sustainability ethics is defined as a consilience or jumping together of econ-ethics and eco-ethics, with both concepts required in the move towards being sustainable (Cairns, 2003). The difference, according to Cairns, lies in the fact the eco-ethics is ecocentric, while sustainability ethics is both ecocentric and homocentric. There is recognition both of the integrated and complex nature of human action, and the need to combine academic disciplines through interdisciplinary research for the purpose of understanding how people impact the planet. Science is invaluable in identifying the endpoints and indicators of how to protect people and the environment, but it is ethics that provides the framework for making sure that these goals are not in violation of the interests of people and the natural world.

3.2.6 Tourism and Animal Ethics

The most recent addition to the literature on tourism and ethics, animal ethics, is one that is perhaps most challenging to embrace. Animals play an important role in the pleasure that tourists derive from zoos, aquaria, dog sledding, greyhound racing, hunting, fishing, ecotourism, cockfighting, badger baiting, bull fighting, rodeos, circuses, and so on. However, by ignoring the animal ethics literature, tourism theorists and practitioners are unable to understand the rightness or wrongness of a range of practices that involve animals.

There are several moral theories that have been useful in navigating this difficult terrain. These include animal rights, utilitarianism, animal welfare, ecocentrism, and ecofeminism (see Fennell, 2012). All provide different perspectives on how or if animals ought to be used for purposes of pleasure and entertainment. For example, animal rights activists argue for empty cages, i.e., animals should never be used for entertainment purposes. The animal welfare platform argues for bigger cages, suggesting that animals can be used if their welfare needs are taken into consideration. Ecocentrism is at odds with the animal rights perspective because it takes into consideration the interests of whole systems and not individual beings.

Recently the use of animal ethics theory in tourism has blossomed. Burns, Macbeth, and Moore (2011) employed ecocentric ethics to evaluate tourism and dingo interactions on Fraser Island, Australia; Fennell and Sheppard (2011) used a number of animal ethics theories to evaluate the sled dog cull after the Vancouver Olympics; Wearing and Jobberns (2011) discussed welfare and ethics in zoos; and Duffy and Moore (2011) examined issues of governance in the welfare considerations of elephants used in the tourism industry. Shani and Pizam (2008) developed an ethical framework for animals used in the tourism industry, with a focus on rights, welfare and environmental ethics. Specific to marine and freshwater systems, Balon used moral philosophy to argue for fishing for subsistence and against fishing for pleasure, especially for tournaments; Hughes (2001) provided a compelling argument against captive dolphin tourism; Garrod (2007) identified a series of ethical issues tied to the use of marine wildlife for tourism; and Fennell and Nowaczek (2010) used biocentrism to argue that fishing should not be viewed as a form of ecotourism.

3.3 Conclusion

There appears to be a certain degree of hesitancy on the part of scholars and practitioners to embrace ethics in tourism. The reasons for this are not entirely clear. It is likely that ethics in tourism suffers from the same problems as business ethics, i.e., how to remain competitive while at the same time maintaining high ethical standards. Scholars who study self-interest and cooperation would argue that people place their own interests above those of others, as well as above priorities tied to the natural world. This is simply human nature (see Fennell, 2006b). The same economic and managerial values are spilling over into education according to Sterling (2001). We are being educated, Sterling writes, to compete and consume rather than to care and conserve. Lost is a sense of authentic education, where caring for the community (broadly conceived) and engagement with real purpose is fleeting: 'People matter, but their attitudes to the natural world and to each other matter most of all' (Duffell, 1998: 78).

In taking this line of thinking one step further, Cairns (2004) writes that Ghandi listed seven deadly social sins: (1) politics without principles, (2) wealth without work, (3) commerce without morality, (4) pleasure without conscience, (5) education without character, (6) science without humanity, and (7) worship without sacrifice. Several of these social sins have relevance to the activities of the tourism industry and to this discussion. Our commercial endeavours must be aided by a sense of what is good or right in human transactions, and the pleasure we derive from touristic activities should not come at a cost in dignity or lifestyle for those

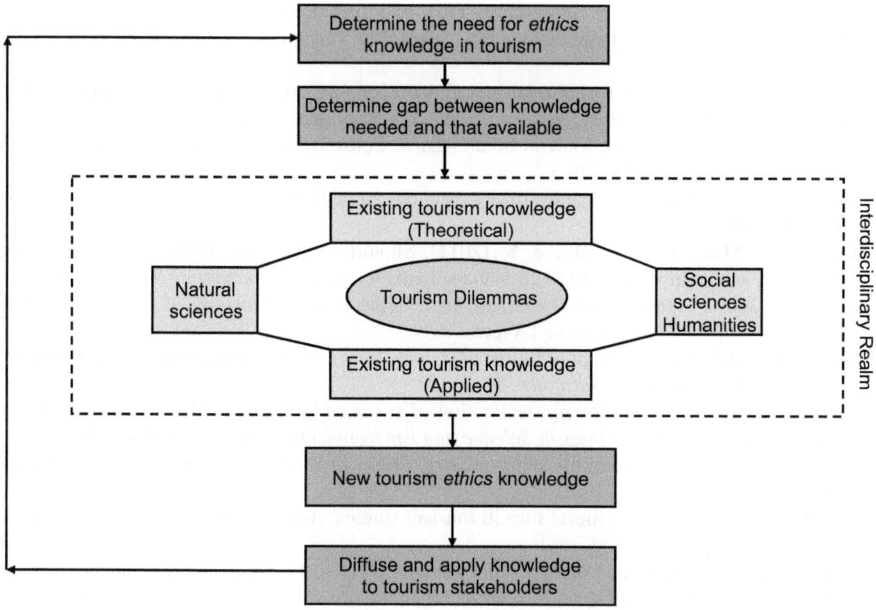

Fig. 3.1 Generating new knowledge in the area of tourism and ethics

marginalised by forces well beyond their control. Education must be firmly entrenched in a narrative that emphasises character development in both teachers and learners, and progress in science must not advance without consideration of the needs of humanity, and the natural world. Figure 3.1 provides a framework for moving the ethics and education agenda forward. It emphasises the importance of interdisciplinary research in creating a tension or disequilibrium in what is presently known in tourism studies. It is only through this foray into other fields of inquiry, that new knowledge can be created and used to pull the tourism field, theoretical and applied, into new and exciting areas of study.

The message here is that ethics provides guidance—a touchstone and a template—on how we ought to think and act in the interests of others, human and nonhuman, in the tourism industry. Proper leadership is required now more than ever from national and international bodies for the purpose of placing tourism within a more rigorous sustainable development agenda. People need to secure lifestyle benefits, but these benefits must be weighed according to present needs and those of the future.

References

Ahmed, Z. U., Krohn, F. B., & Heller, V. L. (1994). International tourism ethics as a way to world understanding. *Journal of Tourism Studies, 5*(2), 36–44.

Baptista, J. A. (2012). The virtuous tourist: Consumption, development, and nongovernmental governance in a Mozambican village. *American Anthropologist, 114*(4), 639–651.

Barber, E. (2011). Case study: Integrating TEFI (Tourism Education Futures Initiative). core values into the undergraduate curriculum. *Journal of Teaching in Travel & Tourism, 11*(1), 38–75.
Baum, T. (2001). Education for tourism in a global economy. In S. Wahab & C. Cooper (Eds.), *Tourism in the age of globalization* (pp. 198–212). London: Routledge.
Boluk, K. A. (2011). Fair trade tourism South Africa: Consumer virtue or moral selving? *Journal of Ecotourism, 10*(3), 235–249.
Brown, L. (2013). Tourism: A catalyst for existential authenticity. *Annals of Tourism Research, 40*, 176–190.
Burns, G. L., Macbeth, J., & Moore, S. (2011). Should dingoes die? Principles for engaging ecocentric ethics in wildlife tourism management. *Journal of Ecotourism, 10*(3), 179–196.
Butcher, J. (2009). Against ethical tourism. In J. Tribe (Ed.), *Philosophical issues in tourism* (pp. 244–260). Clevedon: Channel View.
Butler, R. W. (1980). The concept of tourist area cycle of evolution: Implications for management of resources. *Canadian Geographer, 24*, 5–12.
Cairns, J., Jr. (2003). A preliminary declaration of sustainability ethics: Making peace with the ultimate bioexecutioner. *Ethics in Science and Environmental Politics ESEP*, 43–48.
Cairns, J., Jr. (2004). Sustainability ethics matter. *Ethics in Science and Environmental Politics ESEP*, 3–6.
Caton, K. (2012). Taking the moral turn in tourism studies. *Annals of Tourism Research, 39*(4), 1906–1928.
D'Amore, L. J. (1993). A code of ethics and guidelines for socially and environmentally responsible tourism. *Journal of Travel Research, 31*(3), 64–66.
D'Sa, E. (1999). Wanted: Tourists with a social conscious. *International Journal of Contemporary Hospitality Management, 11*(2/3), 64–68.
Dobson, J. (2011). Towards a utilitarian ethic for marine wildlife tourism. *Tourism in Marine Environments, 7*(3–4), 213–222.
Doxey, G. V. (1975). A causation theory of visitor-resident irritants; methodology and research inference. In TTRA Conference, San Diego, CA (pp. 195–198).
Duffell, R. (1998). Toward the environment and sustainability ethic in engineering education and practice. *Journal of Professional Issues in Engineering Education and Practice, 124*(3), 78–90.
Duffy, R., & Moore, L. (2011). Global regulations and local practices: The politics and governance of animal welfare in elephant tourism. *Journal of Sustainable Tourism, 19*(4–5), 589–604.
Dunfree, T. W., & Black, B. M. (1996). Ethical issues confronting travel agents. *Journal of Business Ethics, 15*, 207–217.
Dyson, P. (2012). Slum tourism: Representing and interpreting 'reality' in Dharavi. *Mumbai, Tourism Geographies: An International Journal of Tourism Space, Place and Environment*. doi:10.1080/14616688.2011.609900.
Eades, J. S. (2009). Moving bodies: The intersections of sex, work, and tourism. In D. C. Wood (Ed.), *Economic development, integration, and morality in Asia and the Americas* (pp. 225–253). Bingley: Emerald.
Enghagen, L. K. (1990). Ethics in hospitality/tourism education: A survey. *Journal of Hospitality & Tourism Research, 14*, 113.
Fennell, D. A. (2006a). *Tourism ethics*. Clevedon: Channel View.
Fennell, D. A. (2006b). Evolution in tourism: The theory of reciprocal altruism and tourist-host interactions. *Current Issues in Tourism, 9*(2), 105–124.
Fennell, D. A. (2008). Responsible tourism: A Kierkegaardian perspective. *Tourism Recreation Research, 33*(1), 3–12.
Fennell, D. A. (2009). The nature of pleasure in pleasure travel. *Tourism Recreation Research, 34* (2), 123–134.
Fennell, D. A. (2012). *Tourism and animal ethics*. London: Routledge.
Fennell, D. A. (2013). The ethics of excellence in tourism research. *Journal of Travel Research, 52* (4), 417–425.

Fennell, D. A., & Malloy, D. C. (1995). Ethics and ecotourism: A comprehensive ethical model. *Journal of Applied Recreation Research, 20*(3), 163–183.

Fennell, D. A., & Malloy, D. C. (2007). *Codes of ethics in tourism: Practice, theory, synthesis.* Clevedon: Channel View.

Fennell, D. A., & Nowaczek, A. (2010). Moral and empirical dimensions of human-animal interactions in ecotourism: Deepening an otherwise shallow pool of debate. *Journal of Ecotourism, 9*(3), 178–189.

Fennell, D. A., & Sheppard, V. (2011). Canada's 2010 Winter Olympic Legacy: Applying an ethical lens to the post-games' sled dog cull. *Journal of Ecotourism, 10*(3), 197–213.

Garrod, B. (2007). Marine wildlife tourism and ethics. In J. Higham & M. Luck (Eds.), *Marine wildlife and tourism management* (pp. 257–271). Wallingford: CABI.

Garrod, B., & Fennell, D. A. (2004). A content analysis of whalewatching codes of conduct. *Annals of Tourism Research, 31*(4), 201–212.

Gibson, C. (2010). Geographies of tourism: (Un)ethical encounters. *Progress in Human Geography, 34*(4), 521–527.

Gretzel, U., Isacsson, A., Matarrita, D., & Wainio, E. (2011). Teaching based on TEFI values: A case study. *Journal of Teaching in Travel & Tourism, 11*(1), 94–106.

Harrison, D. (2008). Pro-poor tourism: A critique. *Third World Quarterly, 29*(5), 851–868.

Heintzman, P. (1995). Leisure, ethics, and the Golden Rule. *Journal of Applied Recreation Research, 20*(3), 203–222.

Higgins-Desbiolles, F. (2006). More than an "industry": The forgotten power of tourism as a social force. *Tourism Management, 27*, 1192–1208.

Higgins-Desbiolles, F. (2008). Justice tourism and alternative globalisation. *Journal of Sustainable Tourism, 16*(3), 345–364.

Hodgkinson, C. (1983). *The philosophy of leadership*. Oxford: Basil Blackwell.

Holcomb, J. L., Upchurch, R. S., & Okumus, F. (2007). Corporate social responsibility: What are top hotel companies reporting? *International Journal of Contemporary Hospitality Management, 19*(6), 461–475.

Holden, A. (1999). High impact tourism: A suitable component of sustainable policy? The case of downhill skiing development at Cairngorm, Scotland. *Journal of Sustainable Tourism, 7*(2), 97–107.

Holden, A. (2003). In need of a new environmental ethics for tourism? *Annals of Tourism Research, 30*(1), 95–108.

Holden, A. (2009). The environment-tourism nexus: Influence of market ethics. *Annals of Tourism Research, 36*(3), 373–389.

Hudson, S., & Miller, G. (2005). The responsible marketing of tourism: The case of Canadian mountain holidays. *Tourism Management, 26*(2), 133–142.

Hudson, S. (2007). To go or not to go? Ethical perspectives on tourism in an 'outpost of tyranny'. *Journal of Business Ethics, 76*(4), 385–396.

Hughes, P. (2001). Animals, values and tourism—Structural shifts in UK dolphin tourism provision. *Tourism Management, 22*(4), 321–329.

Hultsman, J. (1995). Just tourism: An ethical framework. *Annals of Tourism Research, 22*(3), 553–567.

Jamal, T. (2004). Virtue ethics and sustainable tourism pedagogy: Phronesis, principles and practice. *Journal of Sustainable Tourism, 12*(6), 530–545.

Keating, B. (2009). Managing ethics in the tourism supply chain: The case of Chinese travel to Australia. *International Journal of Tourism Research, 11*, 403–408.

Lea, J. P. (1993). Tourism development ethics in the third world. *Annals of Tourism Research, 20*, 701–715.

Lee, S., & Jamal, T. (2008). Environmental justice and environmental equity in tourism: Missing links to sustainability. *Journal of Sustainable Tourism, 7*(1), 44–59.

Lovelock, B. (2008). Ethical travel decisions: Travel agents and human rights. *Annals of Tourism Research, 35*(2), 338–358.

Macbeth, J. (2005). Towards an ethics platform for tourism. *Annals of Tourism Research, 32*(4), 962–984.
Malloy, D. C., & Fennell, D. A. (1998). Codes of ethics and tourism: An exploratory content analysis. *Tourism Management, 19*(5), 453–461.
Mason, P., & Mowforth, M. (2006). Codes of conduct in tourism. *Progress in Tourism and Hospitality Research, 2*(2), 151–167.
McKercher, B. (1993). Some fundamental truths about tourism: Understanding tourism's social and environmental impacts. *Journal of Sustainable Tourism, 1*(1), 6–16.
Meghani, Z. (2011). A robust, particularist ethical assessment of medical tourism. *Developing World Bioethics, 11*(1), 16–29.
Minnaert, L., Maitland, R., & Graham, M. (2006). Social tourism and its ethical foundations. *Tourism, Culture & Communication, 7*(1), 7–17.
Moscardo, G. (2010). Tourism research ethics: Current considerations and future options. In D. G. Pearce & R. Butler (Eds.), *Tourism research: A 20-20 vision* (pp. 203–214). Oxford: Goodfellow.
Moscardo, G., & Murphy, L. (2011). Toward values education in tourism: The challenge of measuring the values. *Journal of Teaching in Travel & Tourism, 11*(1), 76–93.
Nicholls, A. (2010). Fair trade: Towards an economics of virtue. *Journal of Business Ethics, 92*, 241–255.
O'Halloran, R. (1991). Ethics in hospitality and tourism education: The new managers. *Hospitality and Tourism Educator, 3*(3), 33.37.
Payne, D., & Dimanche, F. (1996). Towards a code of conduct for the tourism industry: An ethics model. *Journal of Business Ethics, 15*, 997–1007.
Przeclawski, K. (1996). Deontology of tourism. *Progress in Tourism and Hospitality Research, 2*, 239–245.
Ravinder, R. (2007). Ethical issues in collaboration in the aviation industry. *Tourism Review International, 11*(2), 175–185.
Scheyvens, R. (2007). Exploring the tourism-poverty nexus. *Current Issues in Tourism, 10*(2&3), 231–254.
Selinger, E. (2009). Ethics and poverty tours. *Philosophy & Pubic Policy Quarterly, 29*(1/2), 2–7.
Shani, A., & Pizam, A. (2008). Towards an ethical framework for animal-based attractions. *New Zealand Management, 20*(6), 679–693.
Sheldon, P. J., Fesenmaier, D. R., & Tribe, J. (2011). The Tourism Education Futures Initiative (TEFI), activating change in tourism education. *Journal of Teaching in Travel & Tourism, 11*, 2–23.
Smith, M., & Duffy, R. (2003). *The ethics of tourism development*. London: Routledge.
Speed, C. (2008). Are backpackers ethical tourists? In K. Hannam & I. Atelievic (Eds.), *Backpacker tourism: Concepts and profiles* (pp. 54–81). Clevedon: Channel View.
Sterling, S. (2001). *Sustainable education – Re-visioning learning and change (Schumacher Society Briefing no. 6)*. Dartington: Green Books.
Stonehouse, B. (1997). Tourism codes of conduct in the Arctic and sub-Arctic region. *Journal of Sustainable Tourism, 5*(2), 151–165.
Tribe, J. (2002). The philosophic practitioner. *Annals of Tourism Research, 29*(2), 338–357.
Upchurch, R. S., & Ruhland, S. K. (1995). An analysis of ethical work climate and leadership relationship in lodging operations. *Journal of Travel Research, 34*(2), 36–42.
Wearing, S., & Jobberns, C. (2011). Ecotourism and the commodification of wildlife: Animal welfare and the ethics of zoos. In W. Frost (Ed.), *Zoos and tourism: Conservation, education, entertainment?* (pp. 47–58). Toronto: Channel View.
Weeden, C. (2001). Ethical tourism: An opportunity for competitive advantage? *Journal of Vacation Marketing, 8*(2), 141–153.
Wight, P. A. (1993). Ecotourism: Ethics or eco-sell? *Journal of Travel Research, 21*(3), 3–9.
Williams, P., Gill, A., & Ponsford, I. (2007). Corporate social responsibility at tourism destinations: Toward a social license to operate. *Tourism Review International, 11*(2), 133–144.
Wonders, N. A., & Michalowski, R. (2001). Bodies, borders, and sex tourism in a globalized World: A tale of two cities—Amsterdam and Havana. *Social Problems, 48*(4), 545–571.

Yaman, H. R. (2003). Skinner's naturalism as a paradigm for teaching business ethics: A discussion from tourism. *Teaching Business Ethics, 7*, 107–122.
Yaman, H. R., & Gurel, E. (2006). Ethical ideologies of tourism marketers. *Annals of Tourism Research, 33*(2), 470–489.
Yeung, S. (2004). Hospitality ethics curriculum: An industry perspective. *International Journal of Contemporary Hospitality Management, 16*(4), 253–262.

Chapter 4
Corporate Social Responsibility and Tourism

Dagmar Lund-Durlacher

Those who attend only to the laws of the market should not be surprised if the next generation knows all about monetary values but nothing about moral values
(Johannes Rau, former German President)

Abstract Corporate Social Responsibility (CSR) in tourism can be defined as a guiding business policy whereby tourism companies integrate social and environmental concerns in their own business mission, strategies and operations as well as in their interaction with their stakeholders. As corporations are part of a broader society satisfying stakeholders' needs is central to retaining societal legitimacy and financial liability over the long term. There are different frameworks and guidelines available for integrating CSR into a company's strategy depending on the companies' specific characteristics and their perception of their social responsibilities. CSR today is seen as a multi-stakeholder approach where stakeholders are not only receivers but also partners for realizing and implementing CSR strategies. In tourism companies' strategies the concept of CSR has become a central part. Environmental protection, fair working conditions for employees and contributing to the welfare of local communities are key issues in the strategies of international tourism corporations.

Keywords Corporate Social Responsibility (CSR) • Stakeholder approach • Sustainable Development • Strategy Development • CSR implementation • CSR instruments • CSR reporting

Why do businesses exist? Is their goal simply to maximize profits or do they have other obligations towards the society in which they operate?

D. Lund-Durlacher (✉)
MODUL University Vienna, Vienna, Austria
e-mail: dagmar.lund-durlacher@modul.ac.at

G. Moscardo, P. Benckendorff (eds.), *Education for Sustainability in Tourism*, CSR, Sustainability, Ethics & Governance, DOI 10.1007/978-3-662-47470-9_4

4.1 History and CSR Approaches

There are many past examples of social activism in response to the negative consequences of business activities (e.g. British West Indian slave plantations). Philanthropic entrepreneurs who sought mainly to improve the situation of employees and can be seen as a forerunner of CSR (Werther & Chandler, 2011). The 'modern era' of CSR concepts started in the 1950s when Howard Bowen defined CSR in his book "Responsibilities of a Businessman" as the obligations of companies to reflect the expectations and values of the society in their performance, and thus to envision the total benefit to society as the most important factor for their operations (Bowen, 1953). However, the CSR concept has also received heavy criticism. Friedman (1970) advocated a different approach to CSR in his article "The Social Responsibility of Business is to Increase Profits" where he sees the responsibilities of companies as being limited to creating shareholder value in terms of profit (*Shareholder Approach*). Nevertheless, in the following years, the perceptions of business activities changed and corporations became seen as members of the society which serve the needs of the society and foster social morality in business behavior (*Societal Approach*). Several definitions of CSR subsequently emerged and the CSR concept became more specific (Carroll, 1979, 1991; Davis, 1960, p. 46; Frederick, 1960, p. 60; McGuire, 1963, p. 144; Walton, 1967, p. 18). In 1984, R. Edward Freeman introduced Stakeholder Theory (Freeman, 1984), which states that corporations have relationships with many groups in society (stakeholders) and that responsible corporations must consider the interests of all stakeholders (*Stakeholder Approach*). In the 1990s, related themes such as corporate citizenship, business ethics and corporate governance emerged (for a comprehensive summary of the evolution of the CSR construct, refer to Carroll, 1999).

4.2 Definition

In simple words, CSR is the *business' contribution to sustainable development.* Although there are many deviating definitions and numerous terms used to describe CSR, it is the economic, social and environmental performance, combined with the voluntary nature and the consideration of stakeholder relations which describe the comprehensive scope of CSR. A recent study using frequency counts via a Google search identified the five most common dimensions of CSR: stakeholder dimension, social dimension, economic dimension, voluntariness dimension and environmental dimension (Dahlsrud, 2006 cited in Carroll & Shabana, 2010).

Many CSR definitions are academically derived, but recent contributions have been made by organizations such as the Commission of the European Communities. Following the definition adopted by the European Commission (2011), Corporate Social Responsibility (CSR) can be defined as "*a guiding business policy whereby companies integrate social and environmental concerns in their own business*

Fig. 4.1 The pyramid of corporate social responsibility (Carroll, 1991, p. 42)

mission, strategies and operations and in their interaction with their stakeholders on a voluntary basis". One of the most widely accepted CSR conceptualizations in scientific research was generated by Archie B. Carroll, who presented a CSR pyramid model containing the four categories of Corporate Social Responsibilities (Fig. 4.1), whereby the economic category forms the base on which the legal, ethical and philanthropic categories are built. While businesses should endeavor to perform well in all of the categories at all times, good performance at the lower levels is regarded as prerequisite for achieving the goals of the upper categories (Carroll, 1991).

4.3 Arguments for CSR

Corporations are a part of broader society and interact with different groups within society. It is the society which makes business possible and provides resources such as educated and healthy workers, physical and legal infrastructure, as well as markets for their products. The differences between societies, which have anthropological, sociological, historical and economic dimensions, continuously influence the role of CSR because different societies define the relationship between business and society in different ways. CSR represents an argument for a company's economic self-interest in that satisfying stakeholders' needs is central to retaining societal legitimacy (and therefore financial viability) over the long term (Werther & Chandler, 2011). The *moral argument* for CSR states that corporations' success, besides internal factors, also comes from actions that are congruent with societal values. The *rational argument* for CSR is based on the business' motivation to maximize their performance by minimizing restrictions on operations through actions which go beyond legal requirements and encompass voluntary actions. Societal sanctions (i.e. boycotts, social activism) increasingly impact on corporate success and CSR represents a means of anticipating and reflecting societal concerns to minimize operational and financial limitations on business. The *economic argument* for CSR is founded on the notion that CSR adds value by integrating stakeholders and their needs and thus strengthens the market value of the corporation (Werther & Chandler, 2011). CSR should be dissociated from cause-related marketing, where strategic considerations are missing, and from green-washing where companies promote themselves as acting in an environmentally responsible way without implementing ecological concepts in the company: hence manipulating consumers (a critical discussion on the ethical framework for the marketing of CSR is provided in Van de Ven, 2008).

4.4 Strategy Development and Implementation of CSR

The strategy development and implementation of CSR varies according to corporations' specific characteristics and their perception of their social responsibilities, yet a universally important task is to identify and manage stakeholder expectations (refer also to Matten & Moon, 2008). The Shared Value concept by Michael Porter shifts the emphasis from the perceptions of stakeholders to focus on addressing the prevailing needs of society. Porter and Kramer argue that every "firm should look at decisions and opportunities through the lens of shared value". The concept of shared value can be defined as policies and operating practices that enhance the competitiveness of a company while simultaneously advancing the economies and social conditions in the communities in which it operates (Porter & Kramer, 2011).

Corporate social responsibility strategy development and implementation could be considered an organizational change process and involves learning over time (Maon, Lindgreen, & Swaen, 2009). McElhaney (2009) defines strategic CSR as a "business strategy that is integrated with core business objectives and core competencies of the firm, and from the outset is designed to create business value and positive social change, and is embedded in day-to-day business cultures and operations". Different frameworks and guidelines exist for integrating CSR into strategy. One example is the Environmental-Strategy-Competency-Structure Framework of Werther and Chandler (2011), where strategy is influenced by internal competencies and the external environment (stakeholder expectations) within the boundaries of mission and vision (Werther & Chandler, 2011). McElhaney (2009) suggests a five-step process when creating a CSR strategy:

1. Authentic, firm and public commitment to CSR by management;
2. Develop a CSR strategy that contributes to the top three business objectives of the company;
3. Align CSR strategy with the firms core competencies;

4. Integrate CSR into culture, governance, strategy-development efforts and into existing management and performance systems; and
5. Develop clear performance metrics to measure the impacts of CSR strategies.

It is important that the implementation process includes all levels of corporate decision-making, from top management to front-line personnel. The implementation framework has to be seen as a cycle which consists of six steps: conducting a CSR assessment, developing a CSR strategy, developing CSR commitments, implementing CSR commitments, developing an integrated CSR decision-making structure, verifying and reporting on progress leading to evaluation and improvement. Many governmental and non-governmental organizations provide practical CSR implementation frameworks in their pursuit to promote CSR implementation to corporations (a good example for such implementation frameworks is provided by the International Institute for Sustainable Development (2007).

4.5 Tourism Industry

According to the World Travel and Tourism Council (2013), tourism is both large and developing strongly (2012: 9 % of GDP, 3.2 % growth rate, 101 million jobs). Yet while bringing economic and social benefits to destinations, tourism can also have negative economic, social and environmental impacts (Ap & Crompton, 1998; Archer, Cooper, & Ruhanen, 2005; King, Pizam, & Milman, 1993; Lund-Durlacher, 2013). In 2005, the United Nations World Tourism Organization (UNWTO) together with the United Nations Environment Program (UNEP) defined goals for Sustainable Tourism Development (STD). These goals, based on the triple bottom line approach, refer to economic viability, local prosperity, employment quality, social equity, visitor fulfilment, local control, community well-being, cultural richness, physical integrity, biological diversity, resource efficiency and environmental purity (UNEP and UNWTO, 2005). In 2004, the Tourism Sustainability Group (TSG), launched by the EU Commission in order to provide guidance for the process of STD in Europe, identified eight key challenges to ensuring sustainable tourism development:

- reducing the seasonality of demand,
- addressing the impact of tourism transport,
- improving the quality of tourism jobs,
- maintaining and enhancing community prosperity and quality of life in the face of change,
- minimizing resource use and production of waste,
- conserving and giving value to natural and cultural heritage,
- making holidays available to all, and
- using tourism as a tool in global sustainable development (TSG, 2007).

These areas provide a wide range of opportunities for tourism businesses to engage in STD and to take social responsibility for the natural and social environments in which they operate.

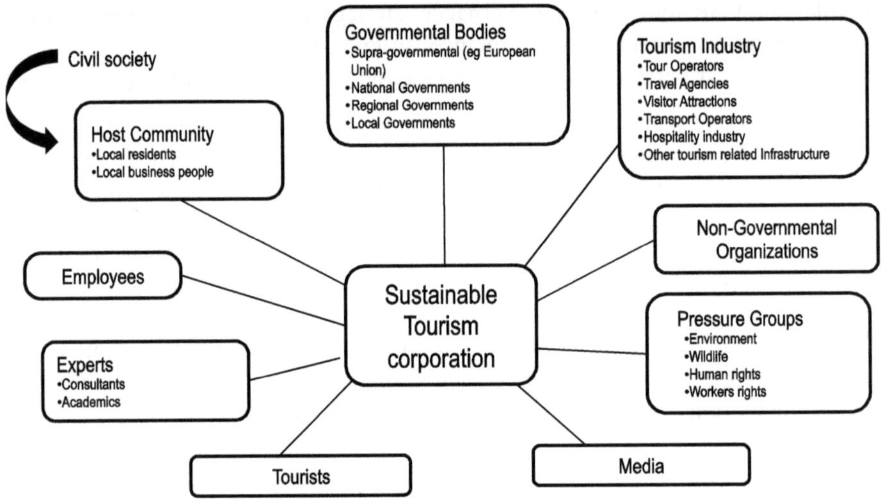

Fig. 4.2 Key stakeholders of tourism businesses (based on Swarbrooke, 2005, p. 17)

As mentioned earlier, CSR is best described as a multi-stakeholder concept where a consideration of the interests of different stakeholders and the dialogue between them plays a major role. According to Freeman, "*a stakeholder in an organization is any group or individual who can affect or is affected by the achievement of the organizations objective*" (Frederick, Post, & Davis, 1992; Freeman, 1984). Important stakeholders for tourism businesses are employees, tourists, other businesses in the supply chain, shareholders, investors, local communities, government authorities, NGOs and the media (Fig. 4.2, based on Swarbrooke, 2005, p. 17). In this context, stakeholders are not only beneficiaries, but also partners for realizing and implementing CSR strategies and projects.

Besides a range of general CSR guidelines, there are a number of basic international strategy papers specifically applicable to tourism companies which may serve as guidelines for businesses to integrate CSR measures:

- The "*Global Code of Ethics for Tourism*" sets a frame of reference for the responsible and sustainable development of tourism (UNWTO, 1999).
- The "*Code of Conduct for the Protection of Children from sexual exploitation in travel and tourism*" developed by ECPAT, a global network of organizations and individuals working together for child protection and signed by many tourism businesses (http://www.thecode.org/). UNWTO has also established a *World Tourism Network on Child Protection* which focuses on protection against the sexual exploitation of minors, child labor and the trafficking in children and young people (http://ethics.unwto.org/en/content/world-tourism-network-child-protection).
- The "*Davos Declaration*" contains adaption and mitigation strategies for tourism to respond to climate change (UNWTO, 2009; UNWTO et al., 2007).

- The "*Tour Operators' Initiative for Sustainable Tourism Development (TOI)*", founded in 2000 as a network of tour operators committed to "developing, operating and marketing tourism in a sustainable manner" and to "making a positive contribution to the natural and cultural environment, which generates benefits for the host communities, and which does not put at risk the future livelihood of local people" (http://www.toinitiative.org).

4.6 CSR Instruments in Tourism

A series of voluntary instruments support businesses to implement CSR practices (Font, 2002; Honey & Stewart, 2002; Toth, 2002). The simplest, but an indispensable means is setting the CSR commitment in the *company vision and mission* as a guiding principle for management, employees and stakeholders.

Codes of conduct and *company CSR programs* are other popular ways of implementing CSR strategies. In the 1990s, several international hotel corporations implemented CSR programs. Examples include Marriott International's program "*Spirit to Serve Our Communities*", and NH Hotels' "*Street Children*". Today, many hotel chains have established their own CSR program and publish annual CSR reports on their websites (e.g. Marriott Hotels, Hilton Hotel Corporation, Inter-Continental Hotels, NH Hotels, Banyan Tree, etc.).

In order to stimulate tourism businesses to implement CSR, a number of *contests* in ecologically and socially responsible tourism activities exist such as the *Tourism for Tomorrow Award*, an annual award presented by the Travel and Tourism Council (WTTC) (http://www.tourismfortomorrow.com/), the "*TO DO!—Contest*" for projects and measures relevant to tourism development whose planning/realization ensure the involvement of the different interests and requirements of local people through participation (www.to-do-contest.org), or "*EcoTrophea*", the international environmental award of the German Travel Association (DRV) (http://www.drv.de/fachthemen/nachhaltigkeit/ecotrophea.html).

More formalized instruments are *Eco- or CSR-management systems* based on international standards such as: the international standard for environmental management in companies (ISO 14001; http://www.iso.org/iso/home/standards/management-standards/iso14000.htm); the international standard for social responsibility (ISO 26000; http://www.iso.org/iso/home/standards/iso26000.htm); or eco-management and audit schemes (such as EMAS; http://ec.europa.eu/environment/emas/index_en.htm).

Since ISO standards and EMAS certification are comprehensive processes and require extensive resources, they are often not suitable for small and medium sized tourism businesses. Besides these comprehensive management schemes, tourism businesses have a number of *certification schemes* and *quality labels for environmentally and socially responsible tourism* available (Bendell & Font, 2004; Honey, 2002). These certification schemes and quality labels are voluntarily adopted by tourism businesses and introduce social responsibility criteria and formalized processes into a corporation's business activities. There are two basic types of certification schemes:

1. Dynamic, process oriented schemes that aim to continuously improve the CSR performance of a company without requiring a certain minimum value to be achieved for target indicators; and
2. Static, result-oriented schemes that measure the achievement of predetermined indicator values.

Today, most certification schemes contain a mix of both approaches. The current landscape of CSR certification schemes and eco-labels for tourism businesses is characterized by a large number of competing schemes and complexity. A large number of eco-labels focused on the accommodation sector are used in Europe. These certification schemes focus mostly on environmental indicators, and systems incorporating aspects of social responsibility are still rare. However this situation is changing as socio-economic and cultural aspects gain greater acceptance as relevant indicators (Font & Buckley, 2001; Font & Epler Wood, 2007; Font & Harris, 2004; Lund-Durlacher, 2013). The Sustainable Tourism Stewardship Council (STSC) was established in 2010 to act as the accreditation body for sustainable tourism certification and eco-labels worldwide. The Global Sustainable Tourism Criteria (GSTC) represent the minimum requirements that hotels and tour operators should apply in order to operate in a sustainable manner (http://www.gstcouncil.org/).

4.7 Effectiveness and Benefits of Certification Schemes

Several studies (Ayuso, 2007; Bader, 2005; Bohdanowicz & Zientara, 2009; Buckley, 2002; El Dief & Font, 2010; Kang, Lee, & Huh, 2010) report on the *effectiveness* and the *benefits* of certification schemes for tourism businesses. According to these studies, the main benefits from the business perspective are cost savings, particularly for water and energy supply, a capacity building process through education and knowledge transfer to management and employees, the implementation of an effective management system and increased employee motivation. The effectiveness as a marketing tool as well as on profitability seems to be limited, although CSR activities seem to have a positive impact on the perceived value of tourism businesses (Kang et al., 2010).

4.8 CSR and Consumers

Customers are increasingly looking for companies that engage in socially responsible activities (Chafe & Honey, 2005). Environmental protection, careful use of natural resources as well as security issues, protection of human rights, social justice and fair working conditions are topics which are of interest to tourists. However, it is important to note that customers pay attention to the motivation of businesses to engage in social activities and only reward a business if the social engagement matches the company's goals and expresses its values, and is not only used as a marketing strategy (Becker-Olsen, Cudmore, & Hill, 2006; Kim, Kang, & Mattila, 2012). While customers are essentially looking for a product or service that satisfies their underlying need, social initiatives can lead to a competitive advantage (Becker-Olsen et al., 2006).

It must also be noted that not all customers are the same. When investigating different types of tourists, it becomes apparent that those interested in nature-based tourism place special importance on responsible innovations; whereas others do consider such actions "important and valuable", but do not focus on them to the same extent (Andereck, 2009). According to Andereck (2009) younger guests put a greater focus on responsible efforts than older people. This is also supported by a survey undertaken and published by The Nielsen Company (2012). According to the survey, younger consumers show a higher willingness to pay more for products and services from socially responsible companies as well as preferring socially responsible companies as workplace or investment opportunities.

4.9 Barriers and Drivers to Implementation of CSR

The main barriers to implementing CSR practices include the amount of time and effort necessary to assess and implement sustainable strategies and practices, and the high investments and operation costs required (Bohdanowicz, 2005). Lack of awareness and knowledge of sustainability among top management as well as scarce governmental support are also barriers for implementing CSR. Tourism businesses are mainly motivated to implement environmentally friendly practices when convinced that they will reduce operational costs and create competitive advantages in the market. Improving image, generating publicity and enabling promotional opportunities are also major desires. Besides these motivations, personal values like the desire for healthy living, as well as awareness and knowledge levels are highly influential for managers considering implementing CSR in their tourism business (Tepelus, 2010).

4.10 CSR Reporting

Concern has been growing about the format and transparency in reporting CSR activities (de Grosbois, 2012; Font, Walmsley, Cogotti, McCombes, & Häusler, 2012). The adoption of reporting standards by the tourism industry, which would facilitate the comparison of different strategies, is still lacking (Bobbin, 2012; Ricaurte, 2011). Bobbin (2012) defines reporting as "a mechanism that enables an organization to manage its impacts on the economy, environment and social conditions by allowing them to measure, track, and improve their performance". Thus, reporting enables a company to create effective CSR strategies, continuously improve their performance and inform their stakeholders about the CSR performance. Today, an increasing number of tourism businesses publish CSR reports which are difficult to compare due to the lack of common reporting standards in the tourism industry. Many tourism businesses base their reporting on the Global Reporting Initiative (GRI) standards (e.g. Accor Hotels, InterContinental Hotels Group, Marriott International and Rezidor Hotel Group). The GRI provides guidelines on how and what to report and is applicable to "organizations of any size, sector, or location" (GRI, 2011). They stress that "*a report should provide a balanced and reasonable representation of the sustainability performance of a reporting organization—including both positive and negative contributions*" (GRI, 2011), while emphasizing the importance of transparent, honest and accurate representation of companies' actions. Currently there are over 3500 organizations reporting according to the GRI guidelines, however only 1.6 % of them are tourism and leisure companies (GRI Database, 2012; cited in Bobbin, 2012).

4.11 CSR in the Hospitality Industry

In the hospitality industry, CSR strategy development and practices are relevant in the following business areas: project planning including site selection, architecture, construction, outside facilities, energy and water supply, disposable systems, housekeeping, food & beverage, programs for (guest) mobility, communications, marketing and customer service points (reservation and reception offices). According to a study among European hoteliers concerning their attitudes towards the environment, 'nearly 85 % of the hoteliers stated that they were involved in some type of environment-oriented activities' (Bohdanowicz, 2005). The main areas of engagement were energy and water conservation and responsible waste management, all leading to significant cost reductions. The spheres of activities relevant to sustainable hospitality management are energy and water management, waste water and waste management, the use of chemicals, contribution to biodiversity and nature conservation, contribution to community development (including employing local people and providing fair and safe working conditions, offering training programs to develop the local labor force, purchasing goods and services from local providers, engaging in cooperation with local providers, supporting social projects to enhance community well-being) and social issues within the work place (including child labor and sexual harassment, fair and equal treatment and fair wages especially for women and indigenous people) (Lund-Durlacher, 2010). Useful manuals and guidelines for implementing CSR practices in the hospitality industry are provided by Sweeting et al. (n.d.), EUHOFA, IHRA, UNEP (2001), UNEP, GTZ (2003). There are many best practice examples of hotels implementing CSR into their strategies and operations; among them the

Planet 21 program of Accor hotels (http://www.accor.com/en/sustainable-development.html) and the *Spirit to Serve our Communities* program of Marriott (http://www.marriott.com/corporate-social-responsibility/corporate-responsibility.mi).

4.12 CSR in Travel Companies

In the past few years, tour operators and travel agencies have become increasingly engaged in implementing CSR measures due to growing consumer awareness and sensibility towards ecologically and socially compliant behavior. Tour operators, which typically combine different travel components such as transportation, accommodation, site visits etc. to create a travel package, face extraordinary challenges when implementing CSR into their business As they not only have to evaluate CSR measures within their own company, but also along the value chain. Recently, a number of CSR certification schemes have been introduced as effective tools to monitor, evaluate and improve CSR practices in tour operator or travel businesses: these include TourCert (www.tourcert.org) and Travelife (http://www.travelife.info/index.php?lang=en). The integration of CSR in the mission statement, as well as key indicators such as financial data, number of employees, consumption of water, power and heat, staff satisfaction and training measures, among others, are relevant CSR aspects concerning these companies. In addition, the core business operations of tour operators are the tours and travels they offer, and these core operations have to be analyzed mainly by evaluating their value chains.

4.13 Future Outlook

In tourism as in other business sectors, the concept of Corporate Social Responsibility has become a central component of companies' strategies. To operate successfully in the future it will be necessary for tourism businesses to continuously implement and realize CSR strategies. Environmental protection, fair working conditions for employees and contributions to the welfare of local communities are key issues in the strategies of international tourism corporations. According to recent publications on CSR trends, the pressure on corporations to develop and implement CSR strategies and practices will increase due to requirements by consumers (sustainability shoppers) and public policies (e.g. reporting obligations, subsidies, taxes). Collaboration across sectors and corporate transparency will also become more important. Trends in the area of corporate social responsibility have recently been identified and published by Mohin (2012) and Visser (2012).

References

Andereck, K. (2009). Tourist's perception of environmentally responsible innovations at tourism business. *Journal of Sustainable Tourism, 17*(4), 489–499.

Ap, J., & Crompton, J. L. (1998). Developing and testing a tourism impact scale. *Journal of Travel Research*. doi:10.1177/004728759803700203.

Archer, B., Cooper, C., & Ruhanen, L. (2005). The positive and negative impacts of tourism. In W. F. Theobald (Ed.), *Global tourism* (pp. 79–102). Burlington, MA: Elsevier.

Ayuso, S. (2007). Comparing voluntary policy instruments for sustainable tourism: The experience of the Spanish hotel sector. *Journal of Sustainable Tourism, 15*(4), 144–159.

Bader, E. (2005). Sustainable hotel business practices. *Journal of Retail & Leisure Property, 5*(1), 70–77.

Becker-Olsen, K. L., Cudmore, B. A., & Hill, R. P. (2006). The impact of perceived corporate social responsibility. *Journal of Business Research, 59*, 46–53.

Bendell, J., & Font, X. (2004). Which tourism rules? Green standards and GATS. *Annals of Tourism Research, 31*(1), 139–156.

Bobbin, J. (2012). Introduction to responsible tourism reporting—The development of a transparent verifiable reporting system suitable for small to medium sized enterprises to monitor their responsible tourism good practices. In *International Centre for Responsible Tourism, Occasional Paper,* OP 23.

Bohdanowicz, P. (2005). European hotelier's environmental attitudes: Greening the business. *Cornell Hotel and Restaurant Administration Quarterly, 46*, 188–204.

Bohdanowicz, P., & Zientara, P. (2009). Hotel companies' contribution to improving the quality of life of local communities and the well-being of their employees. *Tourism and Hospitality Research, 9*, 147–158.

Bowen, H. (1953). *Social responsibilities of the businessman*. New York: Harper and Row.

Buckley, R. C. (2002). Tourism ecolabels. *Annals of Tourism Research, 29*, 183–208.

Carroll, A. B. (1979). A three-dimensional conceptual model of corporate social performance. *Academy of Management Review, 4*, 497–505.

Carroll, A. B. (1991). The pyramid of corporate social responsibility: Toward the moral management of organizational stakeholders. *Business Horizons,* July–August, 39–48.

Carroll, A. B. (1999). Corporate responsibility. Evolution of a definitional construct. *Business & Society, 38*(3), 268–295.

Carroll, A. B., & Shabana, K. M. (2010). The business case for corporate social responsibility: A review of concepts, research and practice. *International Journal of Management Reviews*. doi:10.1111/j.1468-2370.2009.00275.x.

Chafe, Z., & Honey, M. (Eds.) (2005). *Consumer demand and operator support for socially and environmentally responsible tourism* (CESD/TIES Working Paper No. 104). Washington, DC.

Davis, K. (1960). Can business afford to ignore social responsibilities? *California Management Review, 2*, 70–76.

de Grosbois, D. (2012). Corporate social responsibility reporting by the global hotel industry: Commitment, initiatives and performance. *International Journal of Hospitality Management, 31*, 896–905.

El Dief, M., & Font, X. (2010). The determinants of hotels' marketing managers' green marketing behaviour. *Journal of Sustainable Tourism, 18*(2), 157–174.

EUHOFA, IHRA, UNEP. (2001). *Sowing the seeds of change: An environmental teaching pack for the hospitality industry*. Paris: EUHOFA, IHRA, UNEP. http://www.unep.fr/shared/publications/pdf/2991-HospitalityIndustry.pdf. Accessed 3 Sept 2013.

European Commission. (2011). A renewed EU strategy 2011-14 for corporate social responsibility. http://eur-lex.europa.eu/LexUriServ/LexUriServ.do?uri=COM:2011:0681:FIN:EN:PDF. Accessed 4 Sept 2013.

Font, X. (2002). Environmental certification in tourism and hospitality: Progress, process and prospects. *Tourism Management, 23*(4), 197–205.

Font, X., & Buckley, R. (Eds.). (2001). *Tourism ecolabelling: Certification and promotion of sustainable management*. Wallingford: CABI.

Font, X., & Epler Wood, M. (2007). Sustainable tourism certification marketing and its contribution to SME market access. In R. Black & A. Crabtree (Eds.), *Quality assurance and certification in ecotourism* (Ecotourism series, no. 5). Wallingford: CABI.

Font, X., & Harris, C. (2004). Rethinking standards from green to sustainable. *Annals of Tourism Research, 31*(4), 986–1007.

Font, X., Walmsley, A., Cogotti, S., McCombes, L., & Häusler, N. (2012). Corporate social responsibility: The disclosure-performance gap. *Tourism Management, 33*, 1544–1553.

Frederick, W. C. (1960). The growing concern over social responsibility. *California Management Review, 2*, 54–61.

Frederick, W. C., Post, J. E., & Davis, K. (1992). *Business and society—Corporate strategy, public policy, ethics*. New York: McGraw-Hill.

Freeman, R. E. (1984). *Strategic management: A stakeholder approach*. Mansfield, MA: Pitman.

Friedman, M. (1970). The social responsibility of business is to increase its profits. *The New York Magazine*.

Global Reporting Initiative (GRI). (2011). *Sustainability reporting guidelines, version 3.1*. https://www.globalreporting.org/resourcelibrary/G3.1-Sustainability-Reporting-Guidelines.pdf. Accessed 4 Sept 2013.

Honey, M. (Ed.). (2002). *Ecotourism & certification: Setting standards in practice*. Washington: Island Press.

Honey, M., & Stewart, E. (2002). The evolution of green standards for tourism. In M. Honey (Ed.), *Ecotourism & certification: Setting standards in practice* (pp. 33–72). Washington: Island Press.

International Institute for Sustainable Development. (2007). Corporate social responsibility: An implementation guide for business. http://www.iisd.org/pdf/2007/csr_guide.pdf. Accessed 3 Sept 2013.

Kang, K. H., Lee, S., & Huh, C. (2010). Impacts of positive and negative corporate social responsibility activities on company performance in the hospitality industry. *International Journal of Hospitality Management, 29*, 72–82.

Kim, E. E., Kang, J., & Mattila, A. S. (2012). The impact of prevention versus promotion hope on CSR activities. *International Journal of Hospitality Management, 31*, 43–51.

King, B., Pizam, A., & Milman, A. (1993). Social impacts of tourism: Host perceptions. *Annals of Tourism Research, 20*(4), 650–665.

Lund-Durlacher, D. (2010). Hospitality industry. In *Encyclopedia of sustainability: The business of sustainability*, Vol. 2 (pp. 273–275). Great Barrington, MA: Berkshire.

Lund-Durlacher, D. (2013). Corporate social responsibility in tourism. In S. Idowu, N. Capaldi, L. Zu, & A. Das Gupta (Eds.), *Encyclopedia of corporate social responsibility*. Berlin: Springer.

Maon, F., Lindgreen, A., & Swaen, V. (2009). Designing and implementing corporate social responsibility: An integrative framework grounded in theory and practice. *Journal of Business Ethics, 87*(1), 71–89.

Matten, D., & Moon, J. (2008). 'Implicit' and 'explicit' CSR: A conceptual framework for a comparative understanding of corporate social responsibility. *Academy of Management Review, 33*(2), 404–424. http://papers.ssrn.com/sol3/papers.cfm?abstract_id=978942. Accessed 4 Sept 2013.

McElhaney, K. (2009). A strategic approach to corporate social responsibility. Executive forum—Leader to leader. http://responsiblebusiness.haas.berkeley.edu/documents/Strategic%20CSR%20%28Leader%20to%20Leader,%20McElhaney%29.pdf. Accessed 4 Sept 2013.

McGuire, J. (1963). *Business and society*. New York: McGraw-Hill.

Mohin, T. (2012). The top 10 trends in CSR for 2012. http://www.forbes.com/sites/forbesleadershipforum/2012/01/18/the-top-10-trends-in-csr-for-2012/. Accessed 4 Sept 2013.

Porter, M. F., & Kramer, M. R. (2011). Creating shared value. *Harvard Business Review*, January-February. http://www.hks.harvard.edu/m-rcbg/fellows/N_Lovegrove_Study_Group/Session_1/Michael_Porter_Creating_Shared_Value.pdf. Accessed 4 Sept 2013.
Ricaurte, E. (2011). Developing a sustainability measurement framework for hotels: Toward an industry-wide reporting structure. *Cornell Hospitality Report, 11*(13), 6–30.
Swarbrooke, J. (2005). *Sustainable tourism management*. Wallingford: CABI.
Sweeting, J. A. N., & Rosenfeld, A. A. (n.d.). A practical guide to good practice. Managing environmental and social issues in the accommodation sector. http://www.toinitiative.org/fileadmin/docs/publications/HotelGuideEnglish.pdf. Accessed 3 Sept 2013.
Tepelus, C. (2010). Corporate social responsibility in tourism. In J. Liburd & D. Edwards (Eds.), *Understanding the sustainable development of tourism* (pp. 110–129). Oxford: Goodfellow.
The Nielsen Company. (2012). Nielsen. Taken from the global, socially-conscious consumer. http://www.nielsen.com/content/dam/corporate/us/en/reports-downloads/2012-Reports/Nielsen-Global-Social-Responsibility-Report-March-2012.pdf. Accessed 25 Feb 2013.
Toth, R. (2002). Exploring the concepts underlying certification. In M. Honey (Ed.), *Ecotourism & certification: Setting standards in practice* (pp. 73–102). Washington, DC: Island Press.
Tourism Sustainability Group. (2007). Action for more sustainable European tourism: Report of the Tourism Sustainability Group. http://ec.europa.eu/enterprise/sectors/tourism/files/docs/tsg/tsg_final_report_en.pdf. Accessed 4 Sept 2013.
UNEP, GTZ. (2003). A manual for water and waste management: What the tourism industry can do to improve its performance. Paris: UNEP. http://www.unep.fr/shared/publications/pdf/WEBx0015xPA-WaterWaste.pdf. Accessed 3 Sept 2013.
UNEP and UNWTO. (2005). Making tourism more sustainable. A guide for policy makers. http://www.unep.fr/shared/publications/pdf/DTIx0592xPA-TourismPolicyEN.pdf. Accessed 4 Sept 2013.
UNWTO. (1999). Global code of ethics for tourism. http://www.unwto.org/ethics/full_text/en/pdf/CODIGO_PASAPORTE_ING.pdf. Accessed 4 Sept 2013.
UNWTO. (2009). From Davos to Copenhagen and beyond: Advancing tourism's response to climate change. http://www.unwto.org/pdf/From_Davos_to%20Copenhagen_beyond_UNWTOPaper_ElectronicVersion.pdf. Accessed 4 Sept 2013.
UNWTO, et al. (2007). Davos declaration: Climate change and tourism responding to global challenges. http://www.unwto.org/pdf/pr071046.pdf. Accessed 4 Sept 2013.
Van de Ven, B. (2008). An ethical framework for the marketing of corporate social responsibility. *Journal of Business Ethics*. doi:10.1007/s10551-008-9890-1.
Visser, W. (2012). Future trends in CSR: The next 10 years. *CSR International Inspiration Series*, no. 11. http://www.waynevisser.com/wp-content/uploads/2012/12/inspiration_csr_trends_wvisser.pdf. Accessed 4 Sept 2013.
Walton, C. C. (1967). *Corporate social responsibilities*. Belmont, CA: Wadsworth.
Werther, W. B., & Chandler, D. (2011). *Strategic corporate social responsibility: Stakeholders in a global environment* (2nd ed.). Los Angeles: Sage.

Websites

Accor. Planet 21 Program. http://www.accor.com/en/sustainable-development/the-planet-21-program.html. Accessed 4 Sept 2013.
European Commission. Eco-Management and Audit Scheme (EMAS). http://ec.europa.eu/environment/emas/index_en.htm. Accessed 4 Sept 2013.
German Travel Association (DRV), EcoTrophea. http://www.drv.de/index.php?id=545. Accessed 4 Sept 2013.
Global Sustainable Tourism Council (GSTC). http://www.gstcouncil.org/. Accessed 4 Sept 2013.

Institute for Tourism and Development. TO DO!—Contest. http://www.to-do-contest.org/organizer.html. Accessed 4 Sept 2013.
International Organization for Standardization (ISO). ISO 14000—Environmental management. http://www.iso.org/iso/home/standards/management-standards/iso14000.htm. Accessed 4 Sept 2013.
International Organization for Standardization (ISO). ISO 26000—Social responsibility. http://www.iso.org/iso/home/standards/iso26000.htm. Accessed 4 Sept 2013.
The Code. http://www.thecode.org/. Accessed 4 Sept 2013.
Tour Operators Initiative for Sustainable Tourism Development (TOI). http://www.toinitiative.org. Accessed 4 Sept 2013.
Travel and Tourism Council (WTTC). Tourism for Tomorrow Award. http://www.tourismfortomorrow.com. Accessed 4 Sept 2013.
UNWTO World Tourism Network on Child Protection. http://ethics.unwto.org/en/content/world-tourism-network-child-protection. Accessed 4 Sept 2013.

Chapter 5
Tourism and Governance

Dianne Dredge

Abstract This chapter discusses tourism education for sustainability with a particular focus on the challenges and opportunities associated with preparing students to work within complex tourism governance settings. It takes the position that the development of tourism within a sustainability framework requires that tourism professionals effectively engage in dynamic social discourses where difficult trade-offs are made between competing demands. The challenge for tourism education is therefore to prepare graduates to work in these complex, value-laden, socio-political environments where they can proactively and positively contribute to developing forms of tourism that progress the objectives of sustainable development. This chapter explores this challenge in terms of a philosophic tourism practitioner education, and in doing so, discusses three key dimensions of this education: historical antecedents and contemporary knowledge and understandings of governance; competencies for tourism governance for sustainability; and ethical action-oriented practice.

Keywords Tourism • Governance • Philosophic practitioner • Ship of the state • Sustainability • Education

5.1 Introduction

In an increasingly connected world, where governments, business and civil society actors operate within complicated dynamic power sharing arrangements, the capacity to implement sustainable development lies in navigating complex relationships, and in being able to operate effectively to take joint action. Herein lies the challenge of education for sustainability, tourism and governance: tourism education must prepare graduates to work collaboratively with complex multi-scalar problems and to be comfortable with change, uncertainty, ambiguity and competing demands. They must embrace the challenge of addressing the wicked problem of sustainable

D. Dredge (✉)
Aalborg University, AAU-Copenhagen, Denmark
e-mail: dredge@cgs.aau.dk

G. Moscardo, P. Benckendorff (eds.), *Education for Sustainability in Tourism*, CSR, Sustainability, Ethics & Governance, DOI 10.1007/978-3-662-47470-9_5

development with creativity, commitment and an understanding of their own agency. This chapter discusses tourism education for sustainability with a particular focus on the challenges and opportunities associated with preparing students to work within complex tourism governance arrangements.

Contemporary interpretations of governance have established that the concept involves the development and co-ordination of relationships between the state, business and civil society in an effort to 'steer' socio-economic systems (cf. Bramwell, 2011; Hall, 2011; Ladeur, 2004; Pierre, 2000; Rhodes, 1997). In this view, designing, managing and operating governance systems to pursue sustainable development requires much more than a technical education. It requires a philosophic practitioner education inspired by pragmatism and critical management studies. In this philosophic practitioner education students learn to draw from the ideas, theories and tools found in literature and to develop experiential and contextualized knowledge gained from action, discussion, reflection and knowledge sharing. Consequently, a tourism education for sustainability should equip learners with a balance of knowledge, skills and acting abilities so that they can operate in complex social settings to pursue sustainable development (cf. Dredge et al., 2012; Ruwhiu & Cone, 2010; Schon, 1983; Tribe, 2002).

This chapter starts from the viewpoint that sustainable development discourses have historically oversimplified the complex dynamic nature of this challenge. Sustainable development has been treated as an object, an end point or goal, where discussions have taken place without full appreciation of the various on-going and heavily intertwined roles and responsibilities of government, business and civil society (Bramwell & Lane, 2006). This chapter takes the position that the development of tourism within a sustainability framework requires more effective engagement in a dynamic social discourse where difficult trade-offs are made between competing demands (Voss, Bauknnecht, & Kemp, 2006). In other words, sustainable development is not an end point but a dynamic process involving the sharing of knowledge, reflection, communication and the building of trust and mutual respect between actors who have different roles and responsibilities, sources of power, and access to resources. Achieving a type and form of tourism that contributes to sustainable development must therefore involve new forms of dynamic problem framing, and joint action must occur on multiple fronts (Bell & Morse, 2007).

In addressing the challenge of how education can facilitate tourism that contributes to sustainable development, this chapter argues that graduates who will one day take up these challenging roles need to be well versed in three key areas of a philosophic practitioner education:

- Knowledge about governance and sustainable tourism that brings together both local knowledge and global interconnections.
- Practitioner competencies in a range of knowledge building, communication, dispute resolution, capacity building techniques and so on.
- Ethical action-oriented practice that draws from pragmatism and critical reflexive thinking.

Describing and developing these three areas of education is the purpose of this chapter, but before doing so, this chapter will first discuss linkages between education for sustainability, tourism and governance, and make a case for why tourism education must tackle head on the challenge of governance.

5.2 Education for Sustainability, Governance and Tourism

For many tourism researchers and practitioners, discourses that link sustainable development and tourism have made little difference due largely to the simplicity with which the challenge has been treated (Bell & Morse, 2007). Reductionist approaches[1] to studying sustainable development, which often remove the political dimension of the problem (i.e. the existence of multiple interests, competing agendas and power differentials), often have limited value in addressing the practical problems of how to manage tourism. The challenge of sustainable development necessarily involves trade-offs between competing economic, social and environmental priorities, between short and long term outcomes, and between individuals and collectives with varying interests and degrees of power. Education has tended to minimize attention to these complex political dimensions, abstracting the real, difficult and vexed political trade-offs required between social, economic and environmental dimensions. As a result, recommendations become abstract, are 'tacked on' to what are deemed the main findings of research, and are phrased in such a way that governments, business or other organisations *should* do 'this or that'. Such recommendations are often made in a vacuum without appreciating the roles, interests, power and resources that are available and they can be impractical or even irrelevant on the ground. The thorny issue then, is how governments, business and civil society actors can work together to address issues such as social justice and equity and ecological sustainability.

Over the last decade, a range of international organizations and their partners have been working to address this gap, recognizing the importance of and seeking to improve governance in order to facilitate sustainable development (Halle, Najam, & Beaton, 2013; Kemp, Parto, & Gibson, 2005). Drawing from the United Nations' Agenda 21 and the discourses around Education for Sustainability (Wals, 2009), education has a critical role to play in tackling the complexity of the sustainable development challenge in the following ways:

- *Education can promote and improve understandings of sustainable development.* Given that SD is a dialectical concept, and must be interpreted and given

[1] We recognize that reductionist approaches to the study of sustainable tourism management are inevitable given the complexity and interconnectedness of sustainable development problems. It is the inter-connections between studies that require more attention. Moreover, the nature of research funding and academic work in most countries exacerbates the challenge of taking a more integrative approach.

meaning within a context, education can impart the knowledge, skills and perspective necessary to develop locally grounded yet globally connected awareness into the political and value-laden complexity of tourism and sustainable development.

- *Education can help to mobilize individuals and collectives by raising awareness.* Education can help build the capacity of individuals and collectives to share knowledge, raise awareness and make decisions that enhance active and responsible approaches to sustainability.
- *Education can improve the collective capacity of communities to act.* Where people can engage, reflect upon and learn together about SD, a 'learning society' is created that can address sustainability issues collectively to improve societal resilience to sustainability-induced stresses (e.g. climatic events, food security, etc.)

Activities and outputs associated with the United Nations Decade of Education for Sustainable Development (DESD 2004–2015) have highlighted the role of education in finding ways to progress sustainable development. From this work, it becomes clear that alternative ways of thinking, valuing, communicating and acting are needed that allow the myriad of actors involved in making everyday decisions to balance and integrate social, environmental and economic concerns within daily life (Wals, 2009). Here, governance becomes particularly important because it is through effective and coherent governance systems that different actors can come together to discuss, share knowledge, learn, make decisions and implement joint actions to progress sustainable development (Kemp et al., 2005). Sustainable development cannot be achieved without effective governance.

Governance involves the co-ordination of government, business and civil society actors in a process whereby knowledge is shared and actions are identified and implemented to achieve mutually beneficial goals that 'steer' society in a certain direction (Beaumont & Dredge, 2010; Hall, 2011). Given the number of actors involved, effective governance systems require both the creation and maintenance of effective spaces of dialogue, communication and knowledge interchange (i.e. governance processes) and formal administrative bodies that can develop and implement policies and regulatory frameworks (i.e. governance structures). These governance arrangements (both processes and structures) do not just exist but are dynamically created and modified over time by a range of actors involved in and affected by the problem. Issues change, actors move in and out of focus, knowledge flows and actors act based on their interpretation of this information. In this way, effective governance is like a moving target. Governance structures and processes must be both locally appropriate and yet globally engaged. The implications for preparing those who can work in this space to progress sustainability development are, therefore, significant.

In this context, the philosophic practitioner education, much discussed in a variety of professional fields, becomes relevant (cf. Dredge et al., 2012; Marinoff, 2002; Schon, 1983; Schon & Rein, 1994; Tribe, 2002). Without limiting the long line of philosophical thinking that has gone into developing the modern notion of the philosophic practitioner, such a person can be described as a 'stand up philosopher' (Marinoff, 2002), a professional who seeks to facilitate thinking about complex issues drawing upon philosophy as the basis for understanding contemporary problems and identifying concrete actions. It assumes that practitioners have moral agency (in this case to pursue sustainable tourism as a public good), and that this requires a well-developed understanding of values, rights, duties and virtues both of self and others (Helsep, 1997). As such, a philosophical practitioner education draws together three areas of learning:

- Knowledge for governance including both local knowledge and its interconnections with global knowledge.
- Practitioner competencies in a range of knowledge building, communication, dispute resolution, capacity building techniques and so on.
- Ethical action-oriented practice that draws from pragmatism and critical reflexive thinking.

Each of these dimensions will now be examined.

5.3 Knowledge for Tourism, Governance and Sustainability

5.3.1 Historical Development

Governance is not a new term but can be traced back to classical philosophical discussions about who has the power and authority to administer public affairs and control the character (e.g. speed, direction, nature, etc.) of societal change. The etymology of the verb 'to govern' reveals important insights into the meaning of the term. It was Plato's *Republic* (Book VI) where the term *kubernan* was used in a metaphorical exploration of the 'steering' or 'piloting' of the 'ship of the state'.[2] And, while classical philosophy is rarely discussed in tourism education, Plato's imagery is a useful entry point for students to understand the complexity of modern concept of governance.

Plato likens the governance of the city-state to the steering of a ship:

> … there is a captain who is taller and stronger than any of the crew, but he is a little deaf and has a similar infirmity in sight, and his knowledge of navigation is not much better. The sailors are quarrelling with one another about the steering—every one is of opinion that he has a right to steer, though he has never learned the art of navigation (Plato translated by Jowett 2008).

Plato describes the sailors (politicians) as ambitious men, unenlightened by philosophy and who seek to take the helm and steer the ship of the state using the

[2] The term has earlier origins, but Plato's Republic is generally thought to be the first time the term was examined in detail. Later, the Latin verb *gobernare*, 'to direct, rule or guide' was picked up and used in French (*gouverner*) and made its way into Italian and English.

art of persuasion and political strategy (Howland, 2010). The implicit suggestion is that the ship owner should not surrender the helm to sailors ignorant of steersmanship, just like Athenian citizenry should not surrender the leadership of the state to those ignorant of statesmanship (Keyt, 2006). A true steersman or pilot, according to Plato, is one who pays 'attention to year and season and sky and stars and winds and all that belongs to his art' (Rep.VI.488d5-7 in Jowett, 2008). Plato is casting the pilot as someone who is a stargazer who casts his gaze upwards to read the sky and heavens; he is not bothered to look horizontally upon the other sailors to assess their strategies or play their games. In this image, the true pilot is wise and knowledgeable about how to steer the ship; he is not concerned with the other sailors and their unscrupulous competition, neither is he particularly competent in such games. He is viewed as '. . . a babbler and a good-for-nothing by those who sail in ships governed that way'. In this image, Plato suggests that a truly knowledgeable pilot capable of steering society may not be recognized nor valued (he's a good for nothing) by society at large.

While there is much more that can be gained from detailed analysis of the parable of the ship of the state (e.g. see Howland, 2010; Keyt, 2006), Plato's work is useful in our introduction to governance because it alludes to the multitude of competing interests seeking to steer society, and to the ethics and politics that characterize contemporary society. The ship owner (i.e. the citizenry) appear to be preoccupied with the benefits the sailors bestow upon them, while the sailors themselves (politicians) engage in nothing more than quarreling and strategies to dislodge each other from the helm. In such a scenario concern for the broader public interest (i.e. the conduct of trade to strengthen the city-state) is minimized as are all pre-occupied with their own private interests. Also worth reflecting upon at a deeper level, and which we will return to later in the chapter, are the characteristics of the true pilot (a metaphor for the philosopher in the Athenian state), their role in navigating the ship of the state and their relationship with politicians and citizens.

From this classical context, the key questions underpinning the study and teaching of governance are *who* governs, *how* they govern and in *whose interests* is the act of governing (Dredge & Jenkins, 2007; Hall, 2011). Over the years as different philosophical and ideological lenses have been applied, the answers to these questions have changed but the key questions underpinning the term governance remain the same.

5.3.2 Modern Development

For most of the twentieth century, the ideas of Max Weber (1922) and John Maynard Keynes (Keynes, 1936) were heavily influential in questions about *who* governs, *how* they govern and in *whose interests* is the act of governing. Whilst not wishing to limit the importance and wider impact of Keynes' work, in essence government was seen as having a central role intervening in economic affairs to stimulate employment and economic growth. Growing the economy through promoting consumption was the central tenet. Furthermore, through direct government intervention, governments could help to drive economic prosperity which would ultimately have flow-on effects for the well-being of society and serve broader public interests. Under the influence of these ideas tourism was a tool to promote regional economic development, generate foreign exchange and promote employment. Taking a central role, governments invested in tourism infrastructure (airports, roads, ports, protected areas, etc.) in an effort to stimulate further private sector investment in accommodation and attractions (Dredge & Jenkins, 2007).

In another influential stream of thinking, Max Weber's contribution was to argue for the separation of political and bureaucratic arms of government since a professional public service could improve the rigor of government policy-making and make decisions more robust. The contributions of Keynes, Weber and their followers had an enormous influence on the expansion of western bureaucracies over the course of the twentieth century. Bureaucracy expanded both in terms of its size and its policy reach as new policy issues such as tourism and the environment emerged. However, critics of heavy-handed government intervention argued that governments did not have sufficient knowledge of market dynamics and could not be as efficient as the marketplace. As a result, ideas about the role of government in economic management began to change leading to new forms of organization (e.g. public-private partnerships and statutory corporations) and practices (e.g. privatization of public assets, commercialization and outsourcing of services) that were argued to be more efficient (see Dredge & Pforr, 2008). In tourism this was manifested in, for example, the growth of public-private partnerships for destination management and co-funding for the promotion of tourism.

From the 1980s onwards, these shifting roles of government have led to significant changes in the way governments governed. The underpinning tenet of this neoliberal turn, summarized by David Harvey (2005), is that "human well-being can best be advanced by liberating individual entrepreneurial freedoms and skills within a framework characterized by strong private property rights, free markets and free trade". Issues once considered public are now characterised by complex webs of relations between government and non-government interests, and governments must now work collaboratively with non-state actors to manage complex public-private sector issues (Bramwell, 2011).

This refocusing on how governments govern was driven by two main factors. First, driven by the increasing global hegemony of neoliberal economic management, historical notions that governments had 'command and control' and occupy center stage in governing have been replaced with a model of the modern state in which power is shared between public institutions, business and civil society actors (Bramwell, 2006; Krutwaysho & Bramwell, 2010). Second, increasing attention to wicked intractable problems such as climate change, poverty and social justice highlighted that both the power and the responsibility to implement change and move towards sustainable development lies in the capacity of multiple actors to share knowledge, reach mutual understanding and work together in implementing actions (Weber & Khademian, 2008).

Over the last three decades, the term 'governance' has come to denote the co-ordination of government, business and civil society actors in a process whereby shared actions are identified to achieve mutually beneficial goals. As a result of widespread agreement that the answer to '*who* governs' is a complex mix of public-private interests, attention has increasingly shifted to the remaining questions: *how* governance takes place and in *whose interests* is the act of governing. These questions can only be addressed within context because the different institutional settings, the diverse policy actors involved, and the different issues at play generate very different priorities and framings of the sustainable development challenge. We now turn to a discussion of *how* governance takes place in a tourism context, and *whose interests* do these contemporary tourism governance arrangements serve, since these questions are important for the preparation of philosophic practitioners.

5.3.3 Tourism and Governance

Any attempt to implement tourism within a framework of sustainability involves collaboration and joint action, and therefore involves developing and implementing effective governance arrangements (Bramwell, 2007). Effective or 'good' governance arrangements have received considerable attention over the last 10 years, with a number of authors discussing the characteristics of a generic framework of tourism governance that embraces sustainability and that seeks an open and participatory framework to balance social, economic and environmental concerns (e.g. Dredge, 2006; Dredge & Pforr, 2008; Moscardo, 2011). These characteristics suggest that a governance framework for sustainable tourism should be:

- Inclusive of different values and issues and encourage the participation of all individuals and organisations with an interest in sustainable tourism governance
- Consistent with the rule of law
- Transparent in the flow of information and in the way that different interests and power are mediated in decision-making
- Responsive to the widest range of interests
- Oriented towards consensus building and the development of shared understandings and objectives
- Effective in communication and problem solving
- Efficient it its use of resources
- Accountable to the widest range of individuals and groups with an interest both now and in the future.

In practice however, a range of factors complicates the task of developing 'good governance' arrangements. First, tourism interacts and overlaps with a range of other policy areas including, for example, transport, immigration, regional development, environmental management and economic policy. As a result, policies and actions aimed at implementing sustainable tourism must be situated within a broader policy framework of which tourism is only one component. The coordination of actions relies on an integrated approach to sustainable development that has all agencies sharing the same values, agreeing on the same objectives and coordinating their actions in how to get there. Given the fragmentation of the policy space, the dynamics of global-local politics and the flow of both public and private interests, achieving the required level of policy co-ordination is extremely challenging (Bramwell & Lane, 2011; Dredge & Jenkins, 2007).

Second, the influence of the state as the primary driver of policy is declining. Governments have withdrawn from active and direct engagement in policy implementation, seeking instead to use other indirect instruments as a means of achieving their policy aspirations. For example, legal and regulatory instruments to achieve government policy outcomes are being replaced by financial incentive measures (e.g. tax breaks, co-funding programs, etc.), market-led tools (e.g. eco-labelling and accreditation schemes), voluntary guidelines and education programs that rely on private sector support and, quite often, capital investment. The uptake of such measures is reliant on industry capacity, support and good will, which often vary according to economic conditions of the time.

Third, in a case study of the Netherlands (but likely to be applicable in a many countries), Bressers and Dinica (2008) note that while sustainability is 'hot' and tourism is 'booming', sustainable tourism is clearly 'low politics'. A key reason for this observation lies in the policy gaps resulting from a decentralization of policymaking across numerous agencies. The policy challenge of developing tourism that contributes to sustainable development falls into a gap where there is a lack ownership over the problem and a lack of interest in sustainability the beyond financial sustainability of the private sector.

Fourth, in further work Dinica (2009) also notes that despite a generic or symbolic commitment to sustainable tourism development, in practice public agencies take a weaker position on sustainable development because the dominant political ideologies associated with neoliberalism provide a powerful blueprint for the governance of economic sectors such as tourism. This blueprint dictates that industry is better equipped and more knowledgeable about what sort of policy it needs and has come to wield significant power in policy-making. Dredge and Jenkins (2007) have noted similar observations in Australia, as has Bramwell (2011) in the United Kingdom. The problem with this approach of course, is that industry concerns are generally much more short-term and financially focused, and longer-term market failures (e.g. environmental degradation, climate change, rising fuel prices) are not addressed until a (quite often, costly) tipping point emerges.

At an operational level however, the above factors make it difficult to implement a pre-determined or prescribed model of good governance for tourism. Local conditions and contexts mean that such 'good' governance guidelines should be considered aspirational tools to engage with critically and creatively while also employing the lens of local experiences and situated knowledge (cf. Bramwell, 2011; Considine, 2002; Dredge & Pforr, 2008; Grindle, 2008; Klijn & Skelcher, 2007). For example, the capacity of actors to participate and contribute, the type and distribution of knowledge and expertise available, local drivers of development, community aspirations and the balance of power, control and resources to make things happen are all factors that can render prescribed models of good governance for tourism little more than a hollow promise.

The missing link here between the promise of effective governance for sustainable tourism and its implementation lies to some extent in the blending of knowledge, skills and professional capacity of the *in situ* philosophic practitioner. Considering the skills required for such a practitioner is therefore an important aspect of tourism education for sustainable development.

5.4 Competencies for Tourism Governance for Sustainability

The above discussion highlights the need for knowledge about governance to be included in tourism education for sustainability. But knowledge alone is not enough. Building on Plato's parable of the ship of state, being knowledgeable about how to steer (towards sustainability) is only half the challenge, and that skills are also necessary to manage the other stakeholders (e.g. the public, the politicians) so that they too share the same goals and work effectively towards these ends. Here the distinction between skills and competencies comes to the fore: 'skills' signifies proficiency; an aptitude or an ability to undertake a task learned through practice, training and/or experience. 'Competencies' on the other hand denote a set of related abilities that enable a practitioner to undertake a complex job effectively. Charting a course towards sustainability, managing disparate stakeholders while simultaneously assessing and responding to the range of global and local factors that may push the ship off course therefore requires not just a range of skills but a deeper and more robust set of competencies.

A number of authors have identified core competencies in governance for sustainability, and which provide useful insights for tourism. Loorbach (2007), for example, argues that governance for sustainability involves:

- Simultaneously considering different policy domains at multiple levels and in different systems
- Adopting a long term perspective as a framework for short term actions
- Employing a multi-actor approach
- Employing both backcasting and forecasting to reconcile uncertainty
- Embracing pragmatism, critical thinking and reflection

Drawing from discussions in the tourism literature, those involved in tourism governance for sustainable development require a similar range of competencies (e.g. Dredge et al., 2012; Jamal & Menzel, 2009; Tribe, 2002) including, for example:

- Dealing with complexity and uncertainty
- Stakeholder engagement, partnership management and conflict resolution

- Critical thinking, systems and futures thinking
- Action oriented skills to motivate and manage change
- Practical and creative problem solving skills
- Project and process management skills
- Leadership skills

Wiek, Withycombe, and Redman (2011) argue against a 'laundry list' of competencies arguing instead for a conceptually embedded set of interlinked competencies that reflect the problem solving process. In the view of these authors, professionals working in the realm of governance for sustainability should be able to develop, test and implement strategies for sustainable development. In this case, competencies would include:

- Strategic competence—the capacity to identify and steer towards a stronger sustainability position and away from unsustainable trajectories
- Systems thinking competence—capacity to analyse socio-ecological systems, identify leverage or intervention points and assess trade-offs
- Anticipatory competence—the capacity to anticipate, adapt and redirect development trajectories based on an understanding path dependencies and probable causes and consequences
- Normative competence—the capacity to assess alternative strategies and interventions against sustainability criteria
- Interpersonal competence—the capacity to build collaboration, co-produce knowledge and craft a shared vision within diverse stakeholder groups.

Greater engagement with the development of these competencies in education for sustainability will facilitate stronger governance for sustainability.

5.5 Ethical Action Oriented Practice in Tourism Governance for Sustainability

This chapter has so far established that both knowledge of governance and competencies in a range of areas provide important foundations for a philosophic practitioner education. However, even together, they are still not enough. The third dimension of a philosophic practitioner education for tourism governance for sustainability brings together both knowledge and competencies in ethical action-oriented practice.

Since Greek times Aristotle talked of phronesis, a form of practical wisdom that determines how one should act virtuously for a greater good by drawing together knowledge (*episteme*) and technical capabilities (*techne*) (cf. Barnett & Coate, 2005; Dredge et al., 2012; Tribe, 2002). *Episteme* is scientific, universal and context independent knowledge and *techne* is pragmatic craft knowledge—how to do things in a particular context. Phronesis is a 'pragmatic, variable, context-dependent and oriented toward action' type of education (Kinsella & Pitman, 2012, p. 2).

Without limiting the richness of historical discussions around phronesis, in tourism the philosophic practitioner education attempts to capture the ambition of preparing graduates to take an ethical, mindful and engaged role in society. Others go further, prompting social scientists to muster their social and political agency to drive positive change (Dredge & Hales, 2012; Flyvbjerg, 2001; Flyvbjerg, Landman, & Schram, 2012; Hollinshead, Ateljevic, & Ali, 2009). But how to teach ethical action oriented practice—the bringing together of *episteme* and *techne*—is challenging to say the least in contemporary educational settings. Aristotle's world was different to ours, so how can we extract guidance from this ancient concept? At the very least, Aristotle's world was divided into classes and races of people; education and philosophy were the pursuits of the elite; women were precluded from intellectual work; and learning took the form of Socratic discussion (Kinsella & Pitman, 2012). In contrast, our contemporary educational settings are characterised by increasing class sizes and socio-economic diversity and modularized content delivered over a set number of weeks. There is often little opportunity for reflective question-oriented dialogue, and students are more interested in assessment than in learning and reflecting. Education and its role in serving a greater public good is little considered in daily practice although it lurks narrowly in many higher education policies as a tool for economic growth and employment. The question therefore becomes not *how to teach* but *how to create learning opportunities* for students to develop their relationship with the world of tourism, to position themselves in sustainability, and to reflect on their agency and the ethical practitioner they want to be. Proactive agents of governance for sustainability require this foundation.

Dredge, Jenkins, and Whitford (2011) discuss the nature and characteristics of tourism policy making (and by default, governance) and the complex, dynamic context in which tourism graduates will work. For these authors, tourism governance is cast as a value-laden and complex activity that takes place in a variety of fluid policymaking spaces: 'Policymaking takes place within governance and on the edges of organisations; it takes place at rallies and in restaurants, boardrooms and cabinets' (p. 28). These 'small' spaces are where knowledge about tourism and the challenge of sustainable development is co-created and communicated. The philosophic tourism practitioner in tourism governance for sustainability works in these spaces, dynamically and continuously framing, reframing and sharing knowledge about the challenges and potential actions available to implement more sustainable, equitable and just forms of tourism.

For Kemmis (2012), phronesis is a quality of mind, character and action: it develops from the 'sayings, doings and relatings' of practice. From the boarder literature, a slew of adjectives describe such a practitioner: strategic, positive attitude to change, co-operative, entrepreneurial, action-oriented, awareness of and respect for others, awareness of 'boundedness' of one's own thinking and knowledge, flexible, committed and a sense of purpose. Such list of personal qualities, whilst illustrative, provides little guidance for how to deliver learning opportunities to develop the self and one's relationship with the world. For some, work placements and internships scaffolded with reflective learning strategies and assessments, provide promise (e.g. Arendt & Gregoire, 2008; Owusu-Mintah & Kissi, 2012; Wang, Ayres, & Huyton, 2009). However, the extent to which such approaches can connect the learner with their positionality, situatedness, boundedness and potential agency remains unclear and underscores the importance of understanding professional practice by taking a genuine intellectual approach to reflection and action.

Developing an ethical action-oriented dimension to tourism education for sustainability that enables graduates to pursue stronger forms of governance for sustainability must include a range of learning opportunities. Graduates work within a huge range of jurisdictions, in different socio-political, economic and environmental settings, with a range of actors motivated by different interests and capacities, and they address a range of problems of varying complexity. They also bring with them quite different personal experiences, socio-economic and political backgrounds, different disciplinary influences and social networks. As such prescribed approaches to developing this ethical action-oriented dimension to tourism education for sustainability are not appropriate, although a mix of the following formal and informal, *in situ* and classroom learning opportunities have been discussed in literature as having merit:

- Analyzing and reflecting upon existing policy documents in terms of concepts such authorship, issues, interests, power and roles
- Rewriting an existing policy for a different audience or within a different socio-political context
- Responding to a proposed policy from different sectoral or disciplinary perspectives (e.g. from the perspective of an environmental scientist, an elected representative, a community activist, a law enforcement agency, etc.)
- Interviewing policy actors and governance agents about their role, power, interests, strategies, leadership and networks
- Attending and reflecting upon public rallies, meetings and community events
- Facilitating meetings and dispute resolution activities
- Attending and reflecting upon professional networking events.

5.6 Conclusions

This chapter has discussed tourism education for sustainability with a particular focus on the challenges and opportunities associated with preparing students to work within complex tourism governance settings. The challenge is to prepare graduates to work in complex, dynamic, value-laden, socio-political environments, proactively and positively contributing to developing forms of tourism that contribute to a stronger form of sustainable development. To meet this challenge a philosophic practitioner education is required that comprises three dimensions:

- Knowledge about governance and sustainable tourism that brings together both local knowledge and global interconnections.
- Practitioner competencies in a range of knowledge building, communication, dispute resolution, capacity building techniques and so on.
- Ethical action-oriented practice that draws from pragmatism and critical reflexive thinking.

Earlier in this chapter we introduced the parable of the ship of state, explored in Plato's Republic (Book VI). The parable provides useful imagery to help students understand the complex socio-political environment in which they need to work effectively if they are to pursue stronger forms of sustainability. In closing however, we need to point to the inherent danger of taking this or any parable of philosophy at face value. Howland (2010) warns us against assuming that only the helmsman or pilot possesses the overarching wisdom to steer society towards this greater goal. Critics drawing from historical lessons (e.g. the rise of Hitler) warn that when the pilot is vested with the authority and power to steer the ship, they may indeed be motivated by ignoble causes masked as public interest (Howland, 2010). So, while a parable such as the ship of state is useful as a metaphorical entrance for students into the complexity of governance, it is important not to over-privilege its lessons, and to continually return to the parable, teasing it out to understand its strengths and weaknesses. This process itself is highly illustrative and leads to the development of critical thinking and reflection traits discussed above.

The lesson here is that whilst it is important to develop the three dimensions of a philosophic practitioner education discussed in this chapter so that graduates can participate in building stronger forms of governance for sustainability, graduates should not expect that they become the sole arbiters of the public good. Nor are they the only ones to possess the knowledge, competencies and ethical practices to steer tourism governance for sustainability. The practitioner of tourism governance for sustainability must engage with the variety of stakeholders and interests in tourism, and be aware that positive forms of tourism governance for sustainability emerge from social process.

References

Arendt, S. W., & Gregoire, M. B. (2008). Reflection by hospitality management students improves leadership practice scores. *Journal of Hospitality & Tourism Education, 20*(2), 10–15.

Barnett, R., & Coate, K. (2005). *Engaging the curriculum in higher education*. New York: Open University Press.

Beaumont, N., & Dredge, D. (2010). Local tourism governance: A comparison of three network approaches. *Journal of Sustainable Tourism, 18*(1), 7–28.

Bell, S., & Morse, S. (2007). Problem structuring methods: Theorizing the benefits of deconstructing sustainable development projects. *Journal of the Operational Research Society, 58*(5), 576–587.

Bramwell, B. (2006). Actors, power and discourses of growth. *Annals of Tourism Research, 33*(4), 957–978.

Bramwell, B. (2007). Opening up new spaces in the sustainable tourism debate. *Tourism Recreation Research, 32*(1), 1–9.

Bramwell, B. (2011). Governance, the state and sustainable tourism: A political economy approach. *Journal of Sustainable Tourism, 19*(4/5), 459–477.

Bramwell, B., & Lane, B. (2006). Editorial: Policy relevance and sustainable tourism research: Liberal, radical and post-structuralist perspectives. *Journal of Sustainable Tourism, 14*(1), 1–5.

Bramwell, B., & Lane, B. (2011). Critical research on the governance of tourism and sustainability. *Journal of Sustainable Tourism, 19*(4-5), 411–421.

Bressers, H., & Dinica, V. (2008). *Incorporating sustainable development principles in the governance of Dutch domestic tourism: The relevance of boundary judgements.* Paper presented at the international sustainability conference, Basel, 20–22 Aug 2008.

Considine, M. (2002). The end of the line? Accountable governance in the age of networks, partnerships, and joined-up services. *Governance: An International Journal of Policy and Administration, 15*(1), 21–40.

Dinica, V. (2009). Governance for sustainable tourism: A comparison of international and Dutch visions. *Journal of Sustainable Tourism, 17*(5), 583–603.

Dredge, D. (2006). Policy networks and the local organisation of tourism. *Tourism Management, 27*(2), 269–280.

Dredge, D., Benckendorff, P., Day, M., Gross, M. J., Walo, M., Weeks, P., et al. (2012). The philosophic practitioner and the curriculum space. *Annals of Tourism Research, 39*(4), 2154–2176.

Dredge, D., & Hales, R. (2012). Embedded community case study. In L. Dwyer, A. Gill, & N. Seetaram (Eds.), *Handbook of research methods in tourism: Quantitative and qualitative approaches* (pp. 417–437). Cheltenham: Edward Elgar.

Dredge, D., & Jenkins, J. (Eds.). (2007). *Tourism planning and policy*. Milton, QLD: Wiley.

Dredge, D., Jenkins, J., & Whitford, M. (2011). New spaces of tourism planning and policy. In D. Dredge & J. Jenkins (Eds.), *Stories of practice: Tourism planning and policy*. Surrey: Ashgate.

Dredge, D., & Pforr, C. (2008). Tourism policy networks and tourism governance. In N. Scott, R. Baggio, & C. Cooper (Eds.), *Network analysis and tourism* (pp. 58–78). Clevedon: Channel View.

Flyvbjerg, B. (2001). *Making social science matter: Why social inquiry fails and how it can succeed again.* Cambridge: Cambridge University Press.

Flyvbjerg, B., Landman, T., & Schram, S. F. (Eds.). (2012). *Real social science applied phronesis.* Cambridge: Cambridge University Press.

Grindle, M. (2008). *Good governance: The inflation of an idea. JFK School of Government.* Boston, MA: Harvard University.

Hall, C. M. (2011). A typology of governance and its implications for tourism policy analysis. *Journal of Sustainable Tourism, 19*(4&5), 437–457.

Halle, M., Najam, A., & Beaton, C. (2013). *The future of sustainable development: Rethinking sustainable development after Rio+20 and implications for UNEP.* Winnipeg: International Institute for Sustainable Development.

Harvey, D. (2005). *A brief history of neoliberalism.* Oxford: Oxford University Press.

Helsep, R. D. (1997). *Philosophic thinking in educational practice.* Westport, CT: Greenwood.

Hollinshead, K., Ateljevic, I., & Ali, N. (2009). Worldmaking agency—Worldmaking authority: The sovereign constitutive role of tourism. *Tourism Geographies, 11*(4), 427–443.

Howland, J. (2010). Plato's *Republic* and the politics of convalescence. *American Dialetic, 1*(1), 1–17.

Jamal, T., & Menzel, C. (2009). Good actions in tourism. In J. Tribe (Ed.), *Philosophical issues in tourism* (pp. 227–243). Bristol: Channel View.

Kemmis, S. (2012). Phronesis, experience and the primacy of praxis. In E. A. Kinsella & A. Pitman (Eds.), *Phronesis as professional knowledge* (pp. 147–161). Rotterdam: Sense.

Kemp, R., Parto, S., & Gibson, R. B. (2005). Governance for sustainable development: Moving from theory to practice. *International Journal of Sustainable Development, 8*(1), 12–30.
Keynes, J. M. (1936). *General theory of employment, interests and money*. London: Macmillan.
Keyt, D. (2006). Plato and the ship of the state. In G. Santas (Ed.), *The Blackwell guide to Plato's Republic* (pp. 198–213). Laden, MA: Blackwell.
Kinsella, E. A., & Pitman, A. (2012). *Phronesis as professional knowledge*. Rotterdam: Sense.
Klijn, E., & Skelcher, C. (2007). Democracy and governance networks: Compatible or not. *Public Administration, 85*(3), 587–608.
Krutwaysho, O., & Bramwell, B. (2010). Tourism policy implementation and society. *Annals of Tourism Research, 37*(3), 670–691.
Ladeur, K. H. (2004). *Public governance in the age of globalization*. Hants: Ashgate.
Loorbach, D. (2007). Governance for sustainability. *Sustainability: Science, Practice and Policy, 3*(2), 1–4.
Marinoff, L. (2002). *Philosophic practice*. San Diego, CA: Academic Press.
Moscardo, G. (2011). The role of knowledge in good governance for tourism. In E. Laws, E. Richins, J. Agrusa, & N. Scott (Eds.), *Tourist destination governance* (pp. 67–80). Wallingford: CABI.
Owusu-Mintah, S. B., & Kissi, M. (2012). Assessing the effectiveness of internships in tourism education and training in Ghana. *Interdisciplinary Journal of Contemporary Research in Business, 4*(5), 521–540.
Pierre, J. (2000). *Debating governance*. New York: Oxford University Press.
Plato translated by Jowett, B. (2008). *The republic book VI*. http://classics.mit.edu/Plato/republic.7.vi.html. Accessed 28 Dec 2013.
Rhodes, R. A. W. (1997). *Understanding governance: Policy networks, governance, reflexivity and accountability*. Buckingham: Open University Press.
Ruwhiu, D., & Cone, M. (2010). Advancing a pragmatist epistemology in organisational research. *Qualitative Research in Organizations and Management: An International Journal, 5*(2), 108–126.
Schon, D. (1983). *The reflective practitioner*. New York: Basic Books.
Schon, D., & Rein, M. (1994). *Frame reflection: Towards the resolution of intractable policy controversies*. New York: Basic Books.
Tribe, J. (2002). The philosophic practitioner. *Annals of Tourism Research, 29*(2), 228–257.
Voss, J.-P., Bauknnecht, D., & Kemp, R. (2006). *Reflexive governance for sustainable development*. Cheltenham, UK: Edward Elgar.
Wals, A. (2009). *Review of contexts and structures for education for sustainable development 2009*. Paris: UNESCO.
Wang, J., Ayres, H., & Huyton, J. R. (2009). Job ready graduates: A tourism industry perspective. *Journal of Hospitality and Tourism Management, 16*(1), 62–72.
Weber, M. (1922). The essentials of bureaucratic organisations. Reprinted in P. Worsley (Ed.), *Modern sociology* (1978) (pp. 367–371). London: Penguin.
Weber, E., & Khademian, A. (2008). Wicked problems, knowledge challenges and collaborative capacity builders in network settings. *Public Administration Review, 68*(2), 334–349.
Wiek, A., Withycombe, L., & Redman, C. L. (2011). Key competencies in sustainability: A reference framework for academic program development. *Sustainability Science, 6*(2), 203–218.

Part II
Education for Sustainability in Tourism: The Formal Sector

Chapter 6
Sustainable Tourism Education: An Institutional Approach

Dagmar Lund-Durlacher

Abstract Many universities around the globe aim to integrate sustainability into their tourism and hospitality curricula following the need for sustainable tourism development. The goal of this chapter is to discuss how sustainability can be implemented into higher tourism education and what changes are necessary within the university's strategy, organizational structure as well as within the curricula and teaching strategies. Learning for sustainable tourism is not only gaining knowledge about theories related to sustainable tourism but it also calls for changing mind sets and active engagement of the students in matters relating to more sustainable tourism futures. It is concluded that learning also takes place implicitly through the "hidden" curriculum. Staff and educators act as role models for education for sustainable development and students become inspired and motivated by their actions related to sustainability.

Keywords Education • Sustainability • Curriculum development • Holistic education • Values-based education

6.1 Introduction

The goal of this chapter is to discuss and suggest strategies for sustainable tourism education at the university level. The question addressed is whether or not it is now necessary to change or adapt our teaching strategies to be able to teach sustainable tourism at universities and other educational institutions. There are two important aspects to consider when addressing education for sustainable development. First, there are several definitions of the concept of sustainability, but there are still no ideal strategies which would guarantee achieving the optimum state of sustainability. Yet, we still have to explore our tactics towards more sustainable futures as "there is international consensus that achieving sustainable development is essentially a process of learning" (UNESCO, 2002, p. 7). Second, sustainable development "requires a shift in the mental models which frame our thinking and inform

D. Lund-Durlacher (✉)
MODUL University Vienna, Vienna, Austria
e-mail: dagmar.lund-durlacher@modul.ac.at

G. Moscardo, P. Benckendorff (eds.), *Education for Sustainability in Tourism*, CSR, Sustainability, Ethics & Governance, DOI 10.1007/978-3-662-47470-9_6

our decisions and actions" (UNESCO, 2005). It is necessary to question the assumptions on which our current thinking is based, and to contemplate different approaches.

An expert review done under the framework of the UN Decade of Education for Sustainable Development (2004–2015) has identified certain learning processes and key pedagogical approaches to support education for sustainable development. These include "processes which engage the 'whole system', processes of collaboration and dialogues among multi-stakeholders as well as intercultural dialogue, processes which innovate curriculum, as well as teaching and learning experiences and processes of active and participatory learning" (Tilbury, 2011).

6.2 The Sustainability Approach of MODUL University Vienna

In an attempt to explore strategies for implementing sustainable tourism education into the tourism and hospitality curriculum MODUL University Vienna (Austria) will serve as a case study. MODUL University has implemented a sustainability strategy for its undergraduate and (post) graduate tourism and hospitality programs as well as for the university as a whole.

The university is fairly young and was founded in 2007 as a research-oriented international university with undergraduate, graduate, post-graduate and executive programs in the areas of tourism and hospitality management, public governance, sustainable development and management and new media technologies. Sixty international faculty and staff members currently work with and teach about 400 students, representing more than 70 different nationalities and cultural backgrounds. From its inception, the university adopted sustainability as a core value of its mission.

Besides adopting sustainability as a core value, the university decided for a values-based education approach. MU faculty took a leading role in the Tourism Education Futures Initiative (TEFI) and hosted the first TEFI meeting in Vienna in 2007. TEFI "*seeks to provide vision, knowledge and a framework for tourism education programs to promote global citizenship and optimism for a better world*" (TEFI, 2010). At the first TEFI meeting in Vienna participants identified five values-based principles to be embodied in tourism programs for students to become responsible leaders in a fast changing world. Based on TEFI's five value sets ethics, stewardship, knowledge, professionalism and mutuality (TEFI, 2010, p. 9) MODUL University formulated through a long participatory process the four MU educational values of "Knowledge, Creativity & Innovation" (Challenge what we take for granted and embrace change), "Personal Integrity" (Support the principles of equity and justice), "Mutual Respect" (Value diversity and humanity), and "Responsibility and Stewardship" (Serve as ambassador of sustainable and responsible living). These values are introduced to all MU students during orientation week, discussed throughout their studies at MU and at the graduation ceremony students are invited to sign the academic oath and to commit to these values (see Appendix).

Already in the development phase of the university, the development team established cooperation with the Environmental Education FORUM, an initiative of the Austrian Federal Ministry of Agriculture, Forestry, Environment and Water Management and the Austrian Federal Ministry for Education, the Arts and Culture which gained many years of experience in the field of education for sustainable development. Several workshops with members of the university development team as well as representatives of the forum were held in order to identify and define the basic principles of the university's sustainability concept. The discussions were inspired by the sustainability work of the Copernicus Alliance (www.copernicus-alliance.org) as well as other universities with a sustainability focus such as the Eberswalde University for Sustainable Development (www.hnee.de/en), Leuphana University (www.leuphana.de/en/about-us.html), the University of Graz (www.uni-graz.at/en/university/interdisciplinary-affairs/sustainable-university) and the Harvard Green Campus Initiative (www.green.harvard.edu/). Within these workshops, strategies and action plans in the areas administration and management, teaching and curriculum, research and development, structural foundation, communication and decision making, student initiatives and regional integration were developed.

In order to imbed sustainability in teaching and learning the university decided to strategically place sustainability at the university and department level. A mission statement guides the decisions and actions of all academic and administrative units which are fully committed to learning for sustainable development. Deans and program directors were responsible for making decisions regarding the incorporation of sustainability within their educational programs and beyond. Furthermore, a democratic sustainability committee, which serves as an open platform for creative ideas and is responsible for practical implementation of sustainable practices, was established. The committee aims to communicate the idea of becoming a sustainable university to each faculty and staff member as well as to students and other stakeholders of the university and tries to engage all interested stakeholders in a participatory manner.

6.3 The Holistic Approach to Sustainability

An engagement of the whole university system in which the educational programs are based was seen as necessary. This holistic approach means that sustainable tourism education is not only a matter for the curricula or educational programs but also an issue to be implemented within the organizations which facilitate these learning processes. It reflects the notion that educational institutions and teachers are leading by example. In order to teach sustainability, institutions have to consider and reflect on their own practices in the context of sustainability as these indirectly inform teaching and learning. The university has taken diverse measures to improve its sustainability practices including investments in infrastructure such as installing solar panels and a wood pellet heating system.

At a very early stage, MODUL University's sustainability committee put together a sustainability plan which served as the guideline for implementing sustainable practices within the University. Sustainability goals and strategies as well as sustainable practices were developed for the three areas: administration and management, teaching and curriculum and research and development.

For the area of administration and management the sustainability committee has identified six objectives to be focused on in the next stage of the sustainability development whereas the sustainability development has to be understood as a dynamic process which will continuously develop over time and will involve the changing and setting of new goals on a continuous basis as the university develops.

In order to successfully develop, manage and operate a sustainable university the first objective in the sustainability plan which has been derived was "*the establishment of organizational structures for achieving sustainability goals, especially encouraging bottom-up initiatives*". In addition to the sustainability committee, which acts as a steering committee to initiate internal sustainability development projects within the university, a central coordinator for sustainability, who as a member of the sustainability committee is elected by the sustainability committee and responsible for forming workgroups for specific projects and for overseeing them. Further initiatives stimulating bottom-up initiatives are the *Scholarship of Hope* which are yearly donations made to students who come up with innovative, sustainable project ideas and the annual *Sustainability Award* for MU staff and faculty for innovative, sustainable project ideas as well.

The second objective in the sustainability plan was "to manage the operations and facilities of MU in a manner that protects and conserves the environment and supports local businesses, through purchasing environmentally friendly products, locally or regionally produced products, and at the same time reusing and recycling materials and supplies and minimizing waste." Actions such as constantly *raising awareness* of choosing and using environmentally friendly products and supplies (preferably eco-labelled) with decision makers, staff and students, improving *waste separation* by adding waste bins for glass, paper and cans and communicate proper use of bins, selecting a print company that prints CO_2 neutral material, folders, business cards and posters and all other relevant stationery and marketing material, installing electronic devices such as printers, faxes, PCs and lights properly in order avoid wasting energy, setting all printers' default settings to double-sided printing as well as black and white printouts and using rechargeable batteries for all necessary appliances (microphones, laser pointers, presentation remote control, etc.).

Thirdly, the *MU travel policy* as well as the *CO_2 compensation program* was instituted. In order to reduce the carbon footprint, MU adopted and followed a strict travel policy, which is based on the three principles:

1. Avoid travel whenever possible;
2. Use environmentally friendly modes of transportation when appropriate;
3. Compensate CO_2 emissions.

Activities include the use of innovative communication technology to make travel obsolete and encouraging MU staff to use public and environmentally friendly transportation. Based on the criteria of the Austrian environmental organization, flying should be avoided for distances shorter than 700 km; preferably the train should be used. MU only reimburses travel costs arising from public transportation. Furthermore all MU staff flights are compensated through Climate Austria, an Austrian CO_2 compensation provider.

The fourth objective aims to provide a "cafeteria with a healthy, regional and, if possible, organic food and beverage choice to make a contribution to a healthy way of living". The basic principles guiding the cafeteria is to providing fresh and healthy food purchased from preferably local suppliers and produced under fair and environmentally friendly conditions. The cafeteria concept is developed by integrating the views and ideas of all stakeholders including staff, faculty and students.

Objective five bridges the communication gap between the complex issue of sustainability and MU stakeholders and focuses on the "*Establishment of interdisciplinary communication channels, platforms and information structures dealing specifically with issues of sustainable development to provide students, university staff and faculty opportunities to improve their understanding of sustainability and to increase their commitment to act in a sustainable or responsible manner in respect to their regular duties on campus and as global citizen*". All stakeholders should be equally empowered towards making decisions that contribute to their own well-being and to the well-being of the community as a whole. This includes dimensions like education, skills, health, values and leadership. Activities include movie nights for MU students and staff featuring movies on sustainability issues, intercultural days organized by students with the support of MU staff and open to the public, sustainability events dealing with sustainability issues, as well as informing faculty and staff about the actions and strategies of the sustainability committee on a regular basis in the University Assembly (a regular forum consisting of all faculty, staff and student representatives of MU).

The last objective focused on "*Increasing the visibility of MU's sustainability policy and* gaining external recognition as a university pursuing sustainability principles and d*ocumentation of University's sustainability performance*". A visible and tangible symbol of MU's core educational values was installed on the face of the university building displaying the five educational values in 12 different languages to represent the high level of internationality and diversity of students in the MODUL University Vienna community.

The University has participated with four projects in the bi-annual Austrian Sustainability Awards for Universities where all major Austrian Universities participate. Three of the projects were awarded with prices (one project won a first prize). A sustainability report is written each year as part of the University's yearly Activity Report to document the sustainability performance.

When looking at the curriculum of a sustainable tourism and/or hospitality program it is obvious that the concept of sustainability should be included in all courses, at all levels, with linkages between courses. Offering specific sustainability courses equates to treating sustainability as an isolated issue. On the other hand, following an integrated approach, where aspects of sustainability are incorporated in all courses, takes sustainability as an omnipresent topic which permeates all other teaching areas and can be further discussed in different contexts as an underlying philosophy. The curriculum of the BBA for Tourism and Hospitality Management follows this integrated approach encouraging all lecturers to include sustainable tourism issues into each of their courses. Content-wise, the underlying three pillars of sustainability—economic, environmental and socio-cultural issues—and their inter-relationship are addressed.

6.4 Suggested Strategies for Incorporating Sustainability into the Curriculum

Education for sustainable development is strongly aligned with active and participatory learning processes, because "they encourage learners to ask critical reflective questions, clarify values, envision more positive futures, think systemically, respond through applied learning and explore the dialectic between tradition and innovation" (Tilbury, 2011).

Among the tools which have been successfully used in sustainable tourism education are group discussions, which encourage listening and self-reflection, debates for developing arguments, stimulus activities such as watching films or reading a newspaper article in order to stimulate discussion or the use of critical incidents allowing students to reflect their actions on the basis of their moral or ethical attitude.

Case Studies as a research strategy and fieldwork are other popular choices of pedagogy for teaching sustainable tourism (Griffin et al., 2007). They provide the opportunity for students to investigate local issues and work collaboratively with local stakeholders in order to find solutions. Both serve as catalysts for developing students' critical thinking skills in order to understand the complexity of sustainable tourism. Furthermore they can help to influence students' emotions towards a more sustainable development. One example of fieldwork is the involvement of students in a community project with the aim of finding a way to engage community stakeholders in improving CSR practices in a hotel. Through their investigations the students learned about the different perspectives of tourism stakeholders in a particular location, how they were interconnected and what the challenges for sustainable tourism development were. They also developed strong opinions about the necessity of collaboration and this collaboration was seen as the crucial success factor in the sustainable development of the location and the hotel.

Active engagement of the student can also be realized in a virtual environment. Interactive technologies, social networking and the internet provide very important tools for engaging students in learning about and for sustainable tourism. The university participated with a group of undergraduate students in the first Global Online TEFI Courses which were jointly taught by four lecturers based at universities in Denmark, USA, Brazil and Austria (Liburd et al., 2011). While students enjoyed the course and perceived it as interesting, fresh and innovative adding a global dimension, a number of challenges have to be considered including languages, different academic calendars at participating universities and lack of the personal component including feedback loops, support and motivation. Nevertheless, the university participates in the development of the Sustainable Tourism Online Lecture Series, initiated by the BEST (Building Excellence for Sustainable Tourism) Education Network and conducted in partnership with Innotour and several universities around the world in order to provide a global, innovative and collaborative teaching environment to lecturers and students. The online courses consist of a series of online lectures related to the concept of sustainable tourism and students are able to engage actively through blogs and wikis to discuss the course contents and the underlying concepts. It is a collaborative learning approach involving lecturers and students from all over the world.

6.5 Conclusions and Suggestions for Future Strategies

Learning for sustainable tourism development is not only a "process of gaining knowledge, values and theories related to sustainable tourism but it also calls for a changing of mind sets and active engagement of the student in matters relating to more sustainable tourism futures" (Tilbury, 2011). Current curricula and learning environments provided by universities generally do not implement this transformative approach toward education. Learning also takes place implicitly through the "hidden" curriculum. As mentioned earlier, staff and educators act as role models for education for sustainable development and students become inspired and motivated by their actions related to sustainability.

In addition to the integration of sustainability aspects in the curricula courses, the university has introduced a series of measures to stimulate action-taking and sustainable behaviour among their students, faculty and staff, such as the Scholarship of Hope which is awarded to students who propose innovative ideas which contribute to the University's sustainable development, the Sustainability Award for faculty members or the MU Cares program which engages students in social learning. Through the MU Cares program students are stimulated to take responsibility for their student community and beyond, to initiate and engage in charity projects etc. This engagement in social activities contributes to changing their behaviour and changing their mind sets.

This holistic approach in education for sustainable development shows that learning for sustainable tourism is not only a matter of introducing the concept of sustainable tourism into the curriculum, but also about the institutions' and lecturers' commitment towards sustainability, about using teaching approaches which not only create a knowledge base but which also enable social learning toward a sustainable tourism future.

Appendix: Academic Oath

As a MODUL University graduate and a global citizen, I promise:

- I will strive to act with honesty and integrity, oppose all forms of corruption and exploitation, and respect the rights and dignity of all people,
- I will strive to be an ambassador of sustainable and responsible living,
- I will embrace change and creativity to foster progress and innovation,
- I will recognize the knowledge of others, but always challenge what people take for granted.

As I hold true to these principles, it is my hope that I may enjoy an honorable reputation and peace of conscience. This pledge I make freely and upon my honor.

References

Griffin, K., Flanagan, S., Stacey, J., & Tourism Research Centre, DIT. (2007). *Educating for sustainability: Creating a comprehensive, coherent and compelling approach*. Dublin: Fáilte Ireland. Retrieved 10.5.2012, from http://arrow.dit.ie/cgi/viewcontent.cgi?article=1012&context=tfschhmtrep.

Liburd, J. J., Gretzel, U., Lund-Durlacher, D., Sogayar, R. L., Padurean, L., & Schott, C. (2011). TEFI principles in action: Reflections on the first year of the global TEFI courses. In: D. Prebezac (Ed.), *Activating change in tourism education*. TEFI 2011 world congress, congress proceedings (pp. 51–53).

Tilbury, D. (2011). *Education for sustainable development: An expert review of processes and learning*. Paris: UNESCO. Retrieved 10.5.2012, from http://unesdoc.unesco.org/images/0019/001914/191442e.pdf.

Tourism Education Futures Initiative (TEFI). (2010). A values-based framework for tourism education: Building the capacity to lead.

UNESCO. (2002). *Education for sustainability, from Rio to Johannesburg: Lessons learnt from a decade of commitment*. Paris: UNESCO. Retrieved 10.5.2012, from http://unesdoc.unesco.org/images/0012/001271/127100e.pdf.

UNESCO. (2005). *United Nations Decade of Education for Sustainable Development (2005-2014): International implementation scheme*. Paris: UNESCO. Retrieved 10.5.2012, from http://unesdoc.unesco.org/images/0014/001486/148654e.pdf.

Chapter 7
Exploring the Global in Student Assessment and Feedback for Sustainable Tourism Education

Stephen Wearing, Michael A. Tarrant, Stephen Schweinsberg, Kevin Lyons, and Krystina Stoner

Abstract This chapter examines a values-based approach to teaching sustainable tourism management and the related student assessment and feedback mechanisms that reinforce it. The chapter considers and describes how this values-based approach is pedagogically activated by employing critical thinking, self-directed and experiential learning techniques. It draws upon a number of subjects taught by two Universities: one in the USA and one in Australia that use values associated with global citizenship and lifelong learning as frameworks that provide a personally meaningful link between students and the concepts central to sustainable tourism. Particular attention is given to describing alternative assessment and feedback tools that support a values-based approach to sustainable tourism education and to the use of assessable learning contracts, and learning modules.

Keywords Experiential education • Lifelong learning • Sustainability • Global citizenship

7.1 Introduction

In 2012 the United Nations enacted the Higher Education Sustainability Initiative as part of the Rio+20 Summit. This initiative aims, by June 2015, to provide opportunities for 'thousands of university students across multiple disciplines [to gain] knowledge of sustainability concepts that can be applied in the

S. Wearing (✉) • S. Schweinsberg
University of Technology Sydney, Ultimo, NSW, Australia
e-mail: Stephen.Wearing@uts.edu.au; Stephen.Schweinsberg@uts.edu.au

M.A. Tarrant • K. Stoner
University of Georgia, Athens, GA, USA
e-mail: Tarrant@uga.edu; kstoner@uga.edu

K. Lyons
University of Newcastle, Callaghan, NSW, Australia
e-mail: Kevin.Lyons@newcastle.edu.au

G. Moscardo, P. Benckendorff (eds.), *Education for Sustainability in Tourism*, CSR, Sustainability, Ethics & Governance, DOI 10.1007/978-3-662-47470-9_7

marketplace and in living more sustainable lifestyles' (United Nations, 2012). Growth in youth interest in green consumerism (Autio & Heinonen, 2004), along with an increasing alignment of tourism industries globally to Pine and Gilmore's conceptualization of an experience economy has made the youth market who are seeking experiences and green product has made tourism a fertile ground for sustainability scholarship. This trend is evidenced in the development of designated journals (e.g., *Journal of Sustainable Tourism*) and academic think tanks (e.g., BEST Education Network). It is also, unfortunately we would suggest, often too closely connected to what Belhassen and Caton (2011, p. 1389) describe as the alignment of education to 'the logic of capitalist relations of production and consumption'.

Over the past 20 years there has been a formal realization of the value of sustainability education (see Ecologically Sustainable Development Steering Committee, 1992). Schweinsberg, Wearing, and McManus (2013) have noted that such sustainability pedagogy is particularly relevant for the Business Schools in which many tourism educators are now housed. Quigley (2011) has noted that 'in the wake of the global financial crisis, universities were widely condemned for schooling their business graduates in the ruthless pursuit of profit and failing to instil in them a broader set of skills needed to successfully lead business' (p.10).

Of course, not all courses on sustainable tourism are located within business schools. Increasingly, schools of environmental science, geography and forestry are offering such courses. However, the challenge still remains as to how to best prepare students of sustainable tourism in such a way that it can inform their personal and professional lives in the future. To this end, the BEST Education Network has called for tourism students to be equipped with a suite of more generic skills relating to ethics, stewardship, professionalism, knowledge and mutual respect (Sheldon in Liburd & Edwards, 2010, p. viii).

In this chapter we explore how such ambitious goals can be realised through the development and application of a values-based approach to teaching sustainable tourism management. Later in this chapter, we describe and discuss how assessable learning contracts, and learning modules are particularly useful in forging personally meaningful connections to the principles of sustainability as it applies to tourism. However, before exploring these more applied features it is important to describe how a values based approach involves personalizing sustainability. Rather than treating sustainable tourism as a topic to be learned and regurgitated in an examination or essay, a values-based approach asks students to personally engage in experiential learning that is potentially transformative. Such an approach challenges students to consider what it means for them to be global citizens who enact sustainable principles as part of a lifelong learning process.

7.2 Global Citizenship, Lifelong Learning and Sustainable Tourism Principles

A values-based approach to sustainable tourism education is grounded in the value sets associated with becoming a 'good' global citizen. Arguably, in the twenty-first century, the nation is no longer the exclusive framework for social, cultural and political identification (e.g. Banks, 2004). Issues such as climate change, international population flows, cross-border exchanges and the supply and equitable distribution of international resources are undeniably global in scope. Transcending national borders, these issues impact and challenge the very notion of a nationally bounded citizenry (Kofman, 2005). As a result, citizenship is shifting scale, moving away from national affiliations toward global forms of belonging, responsibility and political action (Lyons et al., 2012).

To educate for this it is generally accepted that within these notions exist three key dimensions (Morais & Ogden, 2011; Schattle, 2009; Tarrant et al., 2011): social responsibility, global awareness, and civic responsibility. The authors approach to sustainable tourism education seeks to engage the student in a shift of worldview towards one of global awareness and citizenship. By aligning student assessment and feedback with transformative and reflective experiences, we seek to facilitate students in achieving, fostering and cultivating social good through the deployment of social capital in both host and guest environments. This is a process which has emerged as an important conduit of change and an enabling mechanism in fostering an interconnected global citizenry. An example of this and related assessment techniques as it has been implemented at the University of Georgia are described later in this chapter.

Indeed, the sustainable tourism framework provides a platform by which to engage students as socially responsible citizens (Sheldon, Fesenmaier, & Tribe, 2011; Tarrant, Lyons et al., in press), an educational opportunity that, we believe, is occurring at critical junctures in the students' life course. Encouraging students to grapple with, reframe, and reflect on the 'bigger questions' in this arena (i.e. issues of social justice and equity) will, arguably, impact on young people's life direction, the self-identity of the student, and their relationship to broader social and cultural values and ethics. Results suggest that it is the combination of location (abroad) and academic focus (sustainability) that yields the greatest increases in global awareness and citizenry. Consistent with the values-based approach inherent in the UTS subjects, the conceptual underpinning of Tarrant's model (Wynveen, Kyle, & Tarrant, 2012 an adaptation of Values-Beliefs-Norms theory) is that sustainable tourism promotes a global citizenry by nurturing a sense of justice, pro-environmental behaviours, and civic obligations (Tarrant, 2010).

Consistent with the values-based approach inherent in the UTS subjects, the conceptual underpinning of Tarrant's model (an adaptation of Values-Beliefs-Norms theory) is that sustainable tourism promotes a global citizenry by nurturing a sense of justice, pro-environmental behaviours, and civic obligations (Tarrant, 2010, p. 439). Figure 7.1 suggests that values and worldviews act as filters for new

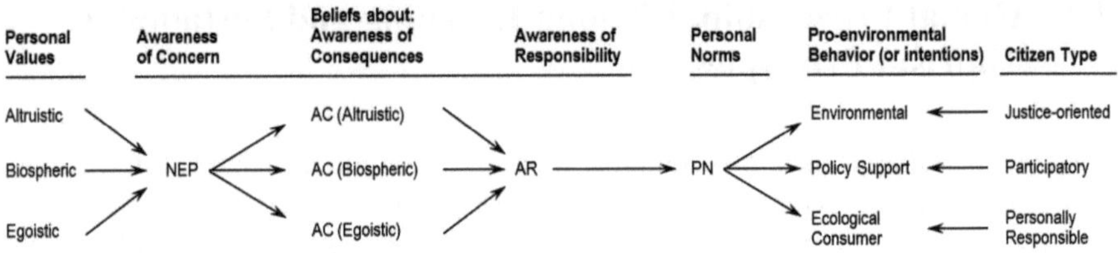

Fig. 7.1 Adapted value-beliefs-norms theory of global citizenship

information in the development and formation of congruent beliefs and attitudes in students which in turn predispose behavioural intentions and ultimately pro-environmental behaviours.

Shepherd (2007) has suggested that sustainability assessments must facilitate the identification of student value positions so that their personal development over time can be monitored. Knowledge of sustainability is a life time process and the critical reflection skills being taught in sustainability units must have relevance beyond the completion of a student's degree, instead laying the foundation for a student to engage in lifelong assessment of complex social situations (Boud, 2000). Blewitt (2004) identifies reflexive life-long learning as one of the principle challenges facing society. Depending on their future career choices, students will precede either into positions where they control the direction of society and its relationship to the natural and social world. Or alternatively they will form part of the wider human society that holds policy makers to account. Whatever their future, students must be provided with the tools and opportunities to drive their own learning processes. This is not to deny the presence of affective skills oriented, learning patterns (Shepherd, 2007). It does, however, call into question the balance between cognitive and affective learning patterns in higher education institutions (Shepherd, 2007).

Effective sustainability education is underpinned by the premise that education must provide intellectual and ethical skills, along with sufficient breadth of industry specific knowledge. Measuring the success of these dual objectives is challenging. While knowledge of tourism industry processes can be measured in exam answers, success of industry internship placements and ultimately job attainment; student appreciation of the intangibles of ethics and stakeholder values will only be illustrated in the way that they respond to ethical challenges in their own business lives. University educators are fortunate to be in a position to lay the groundwork for the development of good corporate citizens. To this end it is important that educators provide opportunities for students to immerse themselves in literature peripheral to their study focus on tourism. Schweinsberg et al. (2013) have proposed that the study of tourism's foundational disciplines (including geography and sociology) may provide the basis for students to appreciate the complexities of tourism's social, economic and environmental relations. Paradigms provide the epistemological lens and access to ways of thinking on subjects such as power and place, which we are argue are essential for ensuring the necessary global awareness to truly appreciate the complexities of the human condition.

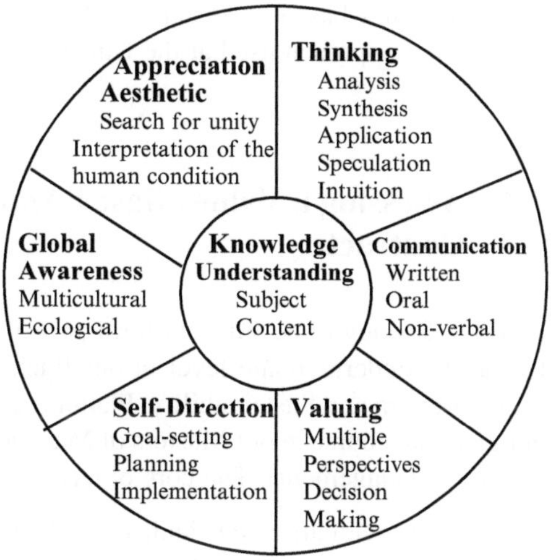

Fig. 7.2 Elements of a tourism education for sustainability approach

Figure 7.2 shows how this can be broken down for the purposes of educational practice; it provides a conceptual view of how elements are related and which components can be grouped.

Blewitt (2004) has identified that sustainability education must be student led, embracive of uncertainty and cross disciplinary based. When students become locked into discipline specific knowledge there is the potential for spontaneity and creativity to be the first casualty of a relentless pursuit of an industry economic imperative. For this reason sustainable tourism educators must work hard to ensure that students recognize the relationship between higher order theoretical discussions and the day to day milieu of a student's practice based education in a business school environment.

Christiansen (nd) suggests that student centered learning is best achieved when traditional power relations are abandoned and the lecturer and student become partners in academic discovery. This is challenging given the tendency for the ever increasingly internationalized student body seeing education simply in terms of passive absorption of data. As Gamache (2007, p. 277) notes; 'many students need an alternative epistemological view, one that enables them to see themselves as creators of 'personal knowledge''. Lea, Stephenson, and Troy (2003, p. 322) identify that student centred learning involves strategically 'increased responsibility and accountability on the part of the student'. Determinations of agreed levels of responsibility start with a learning contract. This contract can introduce the notion of the global along the lines of 'Think globally, act locally' into a practical values based educative approach that relates to the students lifestyle and the effects they are having on their own environments both at home and when they travel. This then leads to relevant ways to assess and provide feedback within this approach.

To illustrate the mechanisms and merits of values-based approaches to student assessment and feedback in sustainable tourism education we refer here to two diverse examples. The first is two subjects at the UTS Business School in Australia where a values-based approach has been adopted. The second is a modularized experiential approach used at the University of Georgia, in the USA.

7.3 Assessing a Values-Based Approach Through Learning Contracts

At UTS, Sydney three subjects form part of our approach to Sustainable Tourism two at the undergraduate level in our Bachelor of Management in Tourism (subjects: Tourism and Sustainability, Planning for Sustainable Destinations) and one in our graduate course in our Masters of Management in Tourism (subject: Sustainable Tourism Management). The core objectives in teaching these units are to:

1. Create individualised curricula of study that embrace their own values, as well as the university's notions of "competence".
2. Consider the definitions of their own professional areas of competence and discuss the knowledge, skills, and attitudes of sustainability as it relates to competent professionals in these fields.
3. Develop proficiency in creating learning contracts as tools for enhancement of self-directed learning about sustainable tourism.
4. Practice and review writing skills; use tools in this manual to maintain a program of continuous improvement.

On successful completion of these sustainability subjects we suggest that participants should be able to:

- Create individualized curricula of study that embrace their own as well as the university's notions of 'competence' in relation to a sustainable tourism related project.
- Define within their professional areas of competence the idea of sustainability through the identification of the knowledge, skills, and attitudes of professionals in their fields through involvement with a sustainable tourism related project.
- Create learning contracts as tools for enhancement of self-directed learning.
- Apply the appropriate skills obtained in a university educational setting to hands-on sustainable community projects.
- Maintain a program of continuous improvement and critical thinking related to sustainability.

The assessment and feedback for these subjects in sustainability is based around a learning contract. At the most fundamental level, learning contracts enable students to engage in a process of collaboration with their teachers to establish agreed upon learning goals (Kearins & Springett, 2003). The contract for the subjects in sustainable tourism management is similar to other contracts in that they are statements of agreements and commitments between two or more parties. However, the approach taken in these subjects is that they are designed to model principles of social sustainability. The processes employed explicitly enacts equitable negotiation between student and teacher, modelling key principles of symmetric two way communication between groups where power might be asymmetrical. Such communication is central to building social capital—a foundation for building social sustainability (Chia, 2011).

The learning contracts developed in these subjects also enable students to develop critical thinking skills that place themselves at the centre of that critique. Rather than a prescriptive set of assessments, the development of a learning contract in these subjects requires students to proactively assess and negotiate customised learning goals that, in part reflect their own career and life goals. Such an approach prepares students for the non-linear career paths that dominate the tourism industry (Lyons, 2010). This resembles what has been described as a portfolio career path, which is strategically crafted by individuals and is based on their perception of the capabilities and attributes they possess in relation to opportunities (Lyons & Brown, 2003).

Central to the task of preparing students for portfolio careers is the establishment of mechanisms whereby students are able to articulate how a particular learning opportunity can contribute to building a skills portfolio which they can access in varying combinations as their lives and careers progress. In order to do this, students need to be given the opportunity to take ownership of their learning goals and these learning contracts provide an ideal tool for this.

In effect, these learning contracts become meaningful to students because of their inherent self-directedness. Students are then motivated to include in their contracts activities and objectives that are specific to their own interest and needs. They then use the learning contract as a means to assess progress toward those objectives.

The first step is a pre-contract worksheet this sets up the subject and learning experiences and to some degree the assessment and feedback mechanisms. The second step looks at learning possibilities, where the student is encouraged to write sentences about what they might learn in the subject and what they might do. Generally we consider it best to begin all their sentences in one of three ways:

1. I'd like to know sustainability and ...
2. I'd like to be able to ...
3. Maybe I will ...

The third step is to select learning objectives. Students are told, 'of the things you said you might like to know or be able to do by the end of the subject, select the ones most important to you.' Having done this, the student can then look at what they said and what they might do to learn these things. They are told that the best method to achieve this is to select those that seem most promising or most interesting, and most likely to help them learn about sustainable tourism and to then list these "activities". It is then time for some feedback and coaching that can be given by the academic. A review of the objectives and activities and then some coaching on how they might be achieved will let the student know whether they are on track, while also providing them with suggestions that might improve what they have and wish to do. Step four is to draft a subject description which gives the reader a brief, general overview of what the subject is about, what areas are to be introduced and covered, what kinds of activities are to be engaged in. It is written in the third person. A good subject description is usually about 100–150 words long (approximately one-half page of close writing on notebook paper or double spaced, typed). The student is encouraged to use their lists of objectives and activities and to write at least 250 words then edit and reduce it to at least 100 but no more than 150 words. They can then compare it to the subject outline and take or send it all (objectives and subject outline) to the subject co-ordinator for review.

Finally, at the end of the subject the student writes a Student Evaluation, a narrative summary of their performance in the subject. It is not part of the contract, but it should be well-connected with the contract. It might include their judgements about how well they feel informed about sustainable tourism, how their completed objectives, matched up to their original intentions with regard to sustainability. The basis for this approach to teaching and in particular assessment and feedback is to provide critical dimensions in this area. As Einstein once said: "We can't solve problems by using the same kind of thinking we used when we created them" (as cited in Milkman, Chugh, & Bazerman, 2009. p. 382). Changes towards a sustainable economy are extremely broad in scope and will require changes in the relationship between people and nature, the relationships between business, government and society (Porter & van der Linde, 1995) and this can only occur if students of sustainability are able to critically evaluate what they see. Without question, these subject contracts could be written more easily, quickly and efficiently. We as teachers could sit down and draft a subject outline in a half-hour or less. But whose subject would it be? Who would be truly responsible for its success? Whose commitment would it have? For the next generation, students need to accept, to some degree, responsibility for what they learn and teaching need to ensure that student have the opportunity to have input into what they learn. Learning Contracts give students the opportunity to construct a learning plan that comes from their own positions, needs, and capabilities. This is the kind of learning plan that is most likely to foster real, significant, and durable learning about sustainability in tourism. Such an outcome is reflected in the following feedback:

> I was quite pleased with what I produced—and actually really enjoyed this assessment! It's surprising how much you actually enjoy doing an assessment when it's something that really interests you. . .This subject is challenging but very beneficial—the aspects of self-directed learning etc. are really great skills to have worked on. (UTS Student Feedback Forms 2009)

7.4 Assessing Modularised Experiential Learning: The University of Georgia Approach

One example of a field subject assessment model can be taken from a suite of University of Georgia (United States) educational travel programs in the South Pacific. These short-term (approximately 4 weeks), experiential programs focus on themes of sustainable development and human-environment relationships (Global Programs in Sustainability, GPS). The GPS programs use field modules as an instructional approach to introducing topics of study. Each module relates to a specific theme and/or geographical location and consists of (a) a background/ introductory narrative, (b) related readings and other associated material (e.g., field activities and instruction, classroom lectures), and (c) several questions (to be answered as discrete essays, peer-review integrative essays, group debates, group projects, field quizzes) about complex ecological, ethical, economic, and social issues relating to sustainability (the "quadruple-bottom line").

The module approach to assessment effectively guides and directs students to construct a 'big picture' of issues relating to sustainable development. Answering each module question requires students to bring together their experiences in the field (which may include service-learning projects, field-based scientific projects, interactions with specialists, and other field activities) along with the material and information from lectures, background readings, group discussions, and individual reflection. In doing this, students actively engage in their own learning process by putting together the various pieces of their experiences to create a holistic view of human relationships with the environment (Tarrant et al., 2011). Indeed, Ewert and Sibthorp (2011) note that a range of confounding variables including precursor, concomitant and post experience conditions may influence the experiential learning process. In the process of creating this global 'big picture', students engage with different stakeholders' perspectives on, and beliefs about, these issues, forcing students to reframe and reconsider their own personal orientations of values and beliefs within the realm of sustainable development.

Such values are assessed through two primary techniques: (1) digital stories and (2) socio-scientific issues (SSI) based simulations. The role of digital stories, as an assessment tool, is described in more detail in a separate chapter of this book and readers are directed to that particular chapter for more information. The focus here will be on SSI's.

For many students, a natural/forested landscape may represent a community of different trees with each species known intimately by one student, or it may simply be a mass of brown and green woods to another student. The former student knows everyone at the party, while the latter may find his or herself alone in a corner knowing no other person. The story to be told, however, does not rely on knowing each species, or indeed even one species, rather it concerns the mosaic of meanings or values represented by the forest. Such values ultimately enable the student to see the forest through a different lens, or worldview, in a way that previously would not have been translated or even considered. Moreover, the values are lived experiences—students form a relationship with the landscape by residing and learning in it—shared amongst one another and with their instructor in both a personal and professional context. The landscape is no longer a collection of trees but a representation of history, politics, and society through the inter-disciplinary module questions that are posed.

In translating this to our students, a primary goal has been to emphasize values—however rudimentary they may be—above pure knowledge. Consistent with the thinking of Aldo Leopold (Leopold, 1949), considered by many to be the Father of Conservation, the importance of simple awareness, appreciation, and sense of interdependence with nature is far greater than the value of knowing the science:

> He who owns a veteran bur oak owns more than a tree. He owns a historical library and a reserved seat in the theater of evolution. . . Education, I fear, is learning to see one thing by going blind to another (Leopold, 1949, p.30).

Leopold's Land Ethic teaches us that unless nature is loved and respected, humans will continue to abuse it as a commodity. Instilling humility is therefore a pre-requisite to establishing a long-term *ethical* relationship that considers the rights of all living species in the community and not solely the dominion of humans. Ethics prompt us to cooperate as members of a community to which we belong and, in an attempt to incorporate ethics into our instruction, we have framed science within a socio-scientific issues (SSI) based approach.

Socio-scientific issues are value-laden and consider the sociocultural and ethical context of real-world scientific problems. In one of the GPS programs, for example, students role-play characters in a simulated local council meeting to develop a deep-water marina in Noosa on Queensland's Sunshine Coast. We have developed a resource package containing political, social, cultural, and economic arguments that reflect the diverse range of opinions from pro- to anti-development. The purpose is to connect students with local decisions that reflect environmental, social, and health issues in their own community. The exercise also seeks to empower students to be active and responsible citizens by equipping them with "the capacity and commitment to take appropriate, responsible, and effective action on matters of social, economic, environmental, and moral-ethical concern" (Hodson, 2011, p. 29).

7.5 Evaluation of a Values Based Approach to Sustainable Tourism

Strapp (1972: 32) sets a number of requirements for environmental education that guide any evaluation of what is learned in sustainable tourism. It includes the following questions:

- Did they get a strong general (global) education, which will help people to develop a 'questioning mind'?
- Do they have an understanding of our natural resources: characteristics, status, distribution and importance to humanity?
- Have they developed an ecological awareness: this being a blend of previous experiences which will develop interest and respect towards the environment?
- Have they developed an economic and political awareness: an understanding of the factors (political and economic) which interfere with conservationist policies?

To achieve this it is necessary to practically engage the student in a reflective critical analysis of their learning about sustainable tourism. One avenue we have found useful is to undertake a subject review through a piece of reflective writing. In psychology and education this is called metacognitive thinking (Flavell, 1979). If you sat down this very minute, and wrote for a minimum of 5 min about what you have learned so far in one of your subjects, that would be like a subject review. Usually, when a person writes for just 5 or 10 min, it is to reflect on the learning accomplished in the previous hour, rather than the past 2 or 3 weeks. When done in an interactive environment where discussion is enabled this reflective analysis provides a platform for reinforcing the importance of values and sustainability.

While we suggest four levels of conducting the subject review, there could be a number of variations. In each case the directions are relatively simple: Ask the student to write about what they considered the most significant elements in what they did and what they learned. Encourage them to talk about what activities held significance, what they read, what they wrote, what they heard and said, and what they felt. This reflective piece should be handed in and reviewed by the subject co-ordinator with the student where possible.

The importance of this evaluation can be seen if one looks at the document *Caring for the Earth* which outlines a plan of action to move towards sustainable living, stating that the benefits and the cost of resource use and environmental conservation should be shared fairly among different communities and it seeks to focus on development as a means of improving the quality of human life while seeing education as a major priority. Within this framework, changing personal attitudes and practices to promote values that support a sustainable way of life is fundamental. and informal and formal education are seen as a means to achieve this, with formal environmental education for children and adults becoming a part of all education levels.

7.6 Conclusions

This chapter outlines how we approach the teaching of sustainability and specifically how assessment and feedback can be used to create a learning environment where the student is offered the opportunity to become self-directed and critical. This perspective is offered as both a part of the University of Georgia Study Abroad program and as a central part of the UTS Business School teaching of Sustainable Tourism. The University of Georgia breaks these learning experiences into a modular process creating this global 'big picture' which is built on smaller interactive learning experiences,

For UTS there is a context where Knight (2010), for example, has reported that business schools are constantly criticised for their lack of social responsibility and accountability of graduates which had an indirect impact on the global financial crisis. There are also positive reasons for change as environmental careers have became more abundant and lucrative. In response to these pressures, a great number of business schools are revising their courses and introducing new subjects on sustainability into their curriculum. The UTS Business School has recently signed the United Nations: Principles for Responsible Management Education (UNPRME) designed to serve as a framework of continuous engagement. It has also cultivated a reputation as a leader in sustainable teaching and was featured in the ARIES report on *Education about and for Sustainability in Australian Business Schools* (Tilbury, Crawley, & Berry, 2004). This context is considered essential to ensure that the delivery of subjects in the area of Sustainable Tourism creates students that are actively engaged. Particularly as sustainability is an inherently vague and complex concept (Phillis & Andriantiatsaholiniaina, 2001) and there is no consensus on its meaning. Its usage in the business community is very loose and has generally tended towards weaker forms of sustainability (Kearins & Springett, 2003). Sustainability has been used to mean different things in a variety of contexts; however, recently sustainability has emerged, not as a concept with some essential meaning, but rather as an important enabling and organizing concept (Miller & O'Leary, 1994). We believe that our approach to teaching is enabling and pragmatic and will hold the student in good stead, and that our means of assessment and feedback builds a student that will engage with the issues of sustainability.

This then meets our responsibility as educators in facilitating change in the society. In the past the role of business education in society has experienced a degree of controversy. Following the Global Financial Crisis there has been considerable discussion on the nature of business education, with particular regard to ethics and corporate governance (Burgan, 2010; Mintzberg, 2010). This builds on two decades of literature critical of business education (Clegg & Ross-Smith, 2003; Mintzberg & Gosling, 2002; Neelankavil, 1994; Pfeffer & Fong, 2002). Welsh and Dehler (2005) point out that there has not been any fundamental change in the models of content nor the process used to educate managers. Our approach seeks to recreate methods that have been established in areas such as environmental education around values based education and adapting them for the area of sustainable tourism education.

The United Nations World Tourism Organization (2013) has estimated global tourism receipts for 2012 at US$1075 billion. The continued profitability of the sector through the Global Financial Crisis necessitates educators treading a fine line between theory and practice (see Jurowski, 2002; Che, 2009). It is not possible for tourism educators to deny their symbiotic relationship to the tourism industry. Neither, however, should educators fall into the trap alleged by Kreisel (2011) where we become beholden to industry masters and engage in pseudo debates and avoid rigorous scientific engagement with the tourism space. One means of avoiding this trap is to ensure the learning environment contains a degree of both critical thinking and individual learning experiences and to link this to student assessment and feedback based on affective outcomes of values, attitudes and behaviours which encourage the valuing of the student and providing experiences in that learning that expose them to circumstances that engage them.

References

Autio, M., & Heinonen, V. (2004). To consume or not to consume? Young people's environmentalism in the affluent Finnish society. *Young Nordic Journal of Youth Research, 21*(2), 137–153.

Banks, J. A. (2004). Teaching for social justice, diversity, and citizenship in a global world. *The Educational Forum, 68*, 289–298.

Belhassen, Y., & Caton, K. (2011). On the need for critical pedagogy in tourism education. *Tourism Management, 32*(6), 1389–1396.

Blewitt, J. (2004). Sustainability and life-long learning. In J. Blewitt & C. Cullingford (Eds.), *The sustainability curriculum: The challenge for higher education*. London: Earthscan.

Boud, D. (2000). Sustainable assessment: Rethinking assessment for the learning society. *Studies in Continuing Education, 22*(2), 151–167.

Burgan, B. (2010). MBA does not spell GFC. *Business Review Weekly, 32*(2), 28–30.

Che, D. (2009). Teaching tourism geography. *Tourism Geographies, 11*(1), 120–123.

Chia, J. (2011). Communicating, connecting and developing social capital for sustainable organisations and their communities. *Australasian Journal of Regional Studies, 17*(3), 330–351.

Christiansen. (n.d.) Premises and practices of discussion teaching. http://hbsp.harvard.edu/multimedia/pcl/pcl_1/resource2/document.pdf. Accessed 10 July 2013.

Clegg, S. R., & Ross-Smith, A. E. (2003). Revising the boundaries: Management education and learning in a postpositivist world. *Academy of Management: Learning and Education, 2*(1), 85–98.

Ecologically Sustainable Development Steering Committee. (1992). National strategy for ecologically sustainable development: Education and training part 3 intersectoral issues—Chapter 26 Challenge. http://www.deh.gov.au/esd/national/nsesd/strategy/educat.html. Accessed 10 July 2012.

Ewert, A., & Sibthorp, J. (2011). Creating outcomes through experiential education: The challenges of confounding variables. *Journal of Experiential Education, 31*(3), 376–389.

Flavell, J. H. (1979). Metacognition and cognitive monitoring: A new area of cognitive-developmental inquiry. *American Psychologist, 34*(10), 906–911.

Gamache, P. (2007). University students as creators of personal knowledge: An alternative epistemological view. *Teaching in Higher Education, 7*(3), 277–294.

Hodson, D. (2011). *Looking to the future: Building a curriculum for social activism*. Rotterdam: Sense.

Jurowski, C. (2002). BEST Thin Tanks and the development of modules for teaching sustainability principles. *Journal of sustainable Tourism, 10*(6), 536–545.

Kearins, K., & Springett, D. (2003). Educating for sustainability: Developing critical skills. *Journal of Management Education, 27*(2), 188–204.

Knight, R. (2010, January 3). A change of track. *Financial Times*. www.ft.com. Accessed 24 Jan 2010.

Kofman, E. (2005). Citizenship, migration and the reassertion of national identity. *Citizenship Studies, 9*(5), 453–467.

Kreisel, W. A. (2011). Some thoughts on the future research on leisure and tourism geography. *Current Issues in Tourism, 15*(4), 397–403.

Lea, S., Stephenson, D., & Troy, J. (2003). Higher education students' attitudes to student centred learning: Beyond 'educational bulimia'. *Studies in Higher Education, 28*(3), 321–334.

Leopold, A. (1949). *A sand county almanac and sketches here and there*. London: Oxford University Press.

Liburd, J., & Edwards, D. (Eds.). (2010). *Understanding the sustainable development of tourism*. Oxford: Goodfellow.

Lyons, K. D. (2010). Room to move? The challenges of career mobility for tourism education. *Journal of Hospitality and Tourism Education., 22*(2), 51–55.

Lyons, K. D., & Brown, P. (2003). Enhancing the employability of Leisure Studies graduates through work integrated learning. *Annals of Leisure Research., 6*(1), 54–67.

Lyons, K., Hanley, J., Wearing, S., & Neil, J. (2012). Gap year volunteer tourism: Myths of global citizenship. *Annals of Tourism Research., 39*(1), 361–378.

Milkman, K. L., Chugh, D., & Bazerman, H. A. (2009). How can decision making be improved? *Perspectives on Psychological Science, 4*, 379–383.

Miller, P., & O'Leary, T. (1994). Accounting, 'economic citizenship' and the spatial ordering of manufacture". *Accounting, Organizations and Society, 19*, 15–44.

Mintzberg, H. (2010). Management education on the fly. Business Week Online 3/9/2010, p. 4. Accessed 3 Nov 2010.

Mintzberg, H., & Gosling, J. (2002). Educating managers beyond borders. *Academy of Management Learning and Education, 1*(1), 64–76.

Morais, D., & Ogden, A. (2011). Initial development and validation of the global citizenship scale. *Journal of Studies in International Education, 15*(5), 445–466.

Neelankavil, J. P. (1994). Corporate America's quest for an ideal MBA. *Journal of Management Development, 15*(5), 38–52.

Pfeffer, J., & Fong, C. T. (2002). The end of business schools? Less success than meets the eye. *Academy of Management: Learning and Education, 1*, 78–96.

Phillis, Y. A., & Andriantiatsaholiniaina, L. A. (2001). Sustainability: An ill-defined concept and its assessment using fuzzy logic. *Ecological Economics, 37*, 435–456.

Porter, M. E., & van der Linde, C. (1995). Toward a new conception of the environment-competitiveness relationship. *Journal of Economic Perspectives, 9*(4), 97–118.

Quigley, R. (2011). Sustaining momentum. *U:Mag, 6*, 10–11.

Schattle, H. (2009). Global citizenship in theory and practice. In R. Lewin (Ed.), *The handbook of practice and research in study abroad : Higher education and the quest for global citizenship* (pp. 3–20). London: Routledge.

Schweinsberg, S., Wearing, S. L., & McManus, P. (2013). Exploring sustainable tourism education in business schools: The honours program. *Journal of Hospitality and Tourism Management, 20*(1), 53–60.

Sheldon, P. J., Fesenmaier, D. R., & Tribe, J. (2011). The Tourism Education Initiative (TEFI): Activating change in tourism education. *Journal of Teaching in Travel & Tourism, 11*(1), 2–23.

Shepherd, K. (2007). Higher education for sustainability: Seeking affective learning outcomes. *International Journal for Sustainability in Higher Education, 9*(1), 87–98.

Strapp, W. B. (1972). Inservice teacher training in environmental education. In William B. Stapp (Ed.). *Environmental education resource book* (pp. 254–260). Newtown, PA: McGraw-Hill.

Tarrant, M. A. (2010). A conceptual framework for exploring the role of studies abroad in nurturing global citizenship. *Journal of Studies in International Education, 14*(5), 433–451.

Tarrant, M. A., Lyons, K., Stoner, L., Kyle, G. T., Wearing, S., & Poudyal, N. (2014). Global citizenry, educational travel, and sustainable tourism: Evidence from Australia and New Zealand. *Journal of Sustainable Tourism.*

Tarrant, M. A., Stoner, L., Borrie, W. T., Kyle, G., Moore, R. L., & Moore, A. (2011). Educational travel and global citizenship. *Journal of Leisure and Research, 43*(3), 403–426.

Tilbury, D., Crawley, C., & Berry, F. (2004). Education about and for sustainability in australian business schools. *Report prepared by the Australian Research Institute in Education for Sustainability* (ARIES) for the Department of the Environment and Heritage, Australian Government. Sydney: ARIES.

United Nations. (2012). Higher Education Sustainability Initiative. http://www.uncsd2012.org/index.php?page=view&type=1006&menu=153&nr=34. Accessed 15 July 2013.

United Nations World Tourism Organisation. (2013). International tourism receipts grew by 4% in 2012. http://media.unwto.org/en/press-release/2013-05-15/international-tourism-receipts-grew-4-2012. Accessed 15 July 2013.

UTS Student Feedback Forms. (2009). University of Technology Sydney, Student Feedback Results, unpublished UTS, Sydney.

Welsh, M. A., & Dehler, G. E. (2005). Whither the MBA? Or the withering of MBAs? *Journal of Studies in International Education, 9*(4), 316–336.

Wynveen, C., Kyle, G. T., & Tarrant, M. A. (2012). Study abroad experiences and global citizenship: Fostering pro-environmental behavior. *Journal of Studies in International Education, 16*, 334–352.

Chapter 8
Cross-Sectorial Relationships for Education for Sustainability

Exploring Innovative Partnerships Between Formal Education and Tourism: Frameworks, Curriculum and Action

Hilary Whitehouse

Abstract The concept of the finite planet underpins all education for sustainability in its current and future forms and iterations. This chapter describes the opportunities available across the formal school and tourism sectors to educate together for a more sustainable means of organizing our lives. International frameworks for environmental education and education for sustainability are described and the Australian frameworks developed in response to the United Nations Decade of Education for Sustainable Development (UNDESD) are explained. Education for sustainability "promotes learning beyond the boundaries of educational institutions" to equip people "with the knowledge, skills and values to address [the] social, environmental and economic challenges of the 21st century" including preparing for jobs that "preserve or restore the quality of the environment" (UNESCO. Five reasons to support ESD—education for sustainable development. http://www.unesco.org/new/index.php?id=96295, 2013). Educators, interpreters and communicators can create networks and relationships for action and learning. There are many variations on this theme of cooperation and accommodation. The formal education sector is porous. The tourism enterprise sector has a huge impact to make in terms of educating and interpreting for sustainability. Aspects of the sustainability cross curriculum priority in the new Australian Curriculum support recalibrating learning practices that enhance sustainability through building partnerships between tourism enterprises and schools. Contemporary policy and curriculum support education for sustainability and the challenge for tourism is to innovate new ways of organizing educational practice. Increasingly we know that cross-sector partnerships can be a highly productive means for learning sustainability. Three examples of practice from far north Queensland reveal how tourism partnerships can be successfully developed with the formal school sector.

H. Whitehouse (✉)
James Cook University, Cairns, QLD, Australia
e-mail: hilary.whitehouse@jcu.edu.au

G. Moscardo, P. Benckendorff (eds.), *Education for Sustainability in Tourism*, CSR, Sustainability, Ethics & Governance, DOI 10.1007/978-3-662-47470-9_8

Keywords Education for sustainability • Frameworks • Australian curriculum • Tourism • Schools • Partnerships

8.1 Introduction

On March 18, 1965, Soviet cosmonaut Alexei Leonov crawled out of an airlock of the Voskhod 2 and walked in space for 12 min. He was so overwhelmed by the view that he shouted the first words he could think of. These were: "The earth is round". At that time, a 'space race' was raging on the technological front of the Cold War between the United States and the Soviet Union. The two superpowers vied with each other to be the first to achieve every major milestone in early space exploration, from sending men on spacewalks to landing men on the moon. (Female astronauts were then excluded despite early experimentation showing female bodies to be comparatively more robust to the demands of space flight.) In one respect, the space race was a battle for conquest characteristic of a world entranced and dominated by Western beliefs of continuous expansion and ideas that the Earth is infinitely exploitable for human gain. But the space race also achieved something completely unexpected—something that put those very beliefs in question: it enabled us for the first time to view our own planet from a distance.

In January 1969, *Life* Magazine, as well as other publications, famously printed "Earthrise," a photo of Earth taken by the astronauts of the historic Apollo 8 mission: an image of a small blue-and-white ball floating in an endless black void. Today we're used to dazzling cosmological images, but a generation ago, the Earthrise image was startling, powerful and unspeakably profound. It gave us, quite literally, the gift of perspective. For the first time, the human population was faced with the reality that our planet is finite. We suddenly understood like never before how alone our world is, how small, how breathtakingly beautiful—and how fragile. In the words of historian Robert Poole (2010, p. 7), "Apollo 8 set out to discover a new world, the moon, and ended up rediscovering its home".

In the years since this discovery of our lonely planet, "sustainability has become the metanarrative of our time" (Blaze Corcoran, 2010). This chapter describes the opportunities available to people across the school and tourism sectors to educate together for a more sustainable means of organizing our lives rather than for the unsustainable conditions in which we find ourselves. The chapter begins with a potted history of the international frameworks that have supported environmental education and education for sustainability in schools before moving on to an explication of the Australian frameworks developed in response to the United Nations Decade of Education for Sustainable Development (UNDESD). Aspects of the sustainability cross curriculum priority in the new Australian Curriculum are introduced before we move on to a discussion of the possibilities for recalibrating our practices to enhance sustainability through building partnerships between tourism enterprises and schools. The frameworks and national curriculum all support education for sustainability across the school, business and community sectors. The challenge is to innovate new ways of organizing educational practice knowing that cross-sector partnerships can be a highly productive means for doing so. Three examples of practice from far north Queensland show how partnerships can be successfully developed and the URL for all relevant documents is included in the text.

8.2 The International Brief for Education for Sustainability: A Potted History

The concept of the finite planet underpins all education for sustainability in its current and future forms and iterations. Education for sustainability began as environmental education with its roots in the post World War II environmental crises and the modern environmental movement that sprang into conscious action to counteract the exponentially increasing destruction of the living fabric of our planet. The scientific community recognized that education was key to both raising awareness and spurring change. The *Australian Academy of Science National Committee for the International Biological Program (IBP) Conference*, held in Canberra in April of 1970, was effectively the first conference on environmental education in Australia. The conference focused on education as significant to increasing community awareness of local environmental degradation, and as necessary to stimulate thought and discussion on social and cultural solutions to problems arising from "the intensification of the interaction between natural and cultural processes".

In the same year, 1970, the *International Working Meeting on Environmental Education in the School Curriculum* was convened by the International Union for the Conservation of Nature and Natural Resources (IUCN) in Nevada, USA. Environmental education was then constructed as involving the teaching of skills, attitudes, values, decision-making and practical action. Also in 1970, the United States enacted legislation called *The Environmental Education Act*, which defined environmental education as "intended to promote among citizens the awareness and understanding of the environment—our relationship to it and the concern and possible action necessary to assure our survival and improve the quality of life". This novel piece of American legislation noted the purpose of environmental education was ameliorative. That is, the field was conceived as having instrumental value. Its stated educational purpose was to attempt to alter the ways in which consumer societies were behaving. The language was dramatic (and continues to be so all these years later). No other educational discipline field has been given the explicit task of assuring human survival.

The United Nations has always played a significant role in defining the international parameters of education for sustainability. In 1977, the UNESCO-UNEP *Intergovernmental Conference on Environmental Education* was held in Tbilisi, Georgia. This produced the historically significant document known as the *Tbilisi*

Declaration that defined the major objectives of environmental education as awareness, knowledge, attitudes, skills and participation, which is a critical, distinguishing feature. Participatory action is political, which is one reason why environmental education has been continually marginalized in conservative school systems (see Gough, 1997). Both state and private schools generally try to be apolitical even as they reproduce societal relations (including political relations) within their own structures. Environmentalism has and still is positioned as radical practice, as counter to the business-as-usual (BAU) conduct of ecologically destructive economic relations. This is why the action component of the *Tbilisi Declaration* has always been so important for environmental educators and sustainability educators.

In 1987, the United Nations published the impressively detailed *Brundtland Report* that established the concept of sustainability as meeting the needs of the present without sacrificing the rights and needs of those in the future. This led to national governments meeting at the 1992 *United Nations Conference on Environment and Development (UNCED) in Rio de Janeiro* (known as the first Earth Summit) where the attendees produced a voluntary action plan called *Agenda 21* and 178 nations (including the USA and Australia) eventually signed up. *Agenda 21* presented a vision for how all levels of governments (local, state and national) can take action to combat poverty and pollution, conserve biodiversity and natural resources and develop their populations in a sustainable manner. *Agenda 21* is very popular on the African continent but it hasn't been smooth sailing in the U.S.A. Many American states, including Mississippi, Kansas and Tennessee, have tried to pass legislation banning *Agenda 21* as a global conspiracy to deny people constitutional property rights. But *Agenda 21* is not an international treaty. It is a voluntary, global action plan setting desired directions for a more pleasant (finite) world, or as some have seen it, as a plan to stave off Armageddon.

By 2002, governments were meeting again at the *United Nations World Summit on Sustainable Development* (WSSD), also known as Rio + 10 and Earth Summit 2002, held in Johannesburg, South Africa. The major talking point was that nations had made unacceptably sluggish progress toward the education goals set forth 10 years earlier. *Agenda 21* had called for all countries to implement an education for sustainable development strategy by 2002. At WSSD, education and learning were again endorsed as a key means for capacity building sustainable futures. The Japanese Government conceived of a plan to be implemented over a decade and with massive lobbying from governmental and non-governmental organizations, the United Nations Decade of Education for Sustainable Development was endorsed. As UNESCO described it, the UNDESD was "an educational effort" "to integrate the principles, values, and practices of sustainable development into all aspects of education and learning, in order to address the social, economic, cultural and environmental problems we face in the 21st century".

The problematic terminology of "education for sustainable development" and the vagueness of the definition fuelled a heated academic debate concerning issues of the etymology, pedagogy, and concept of sustainable development. Some academics argued that ESD adopts an anthropocentric perspective of the biosphere, viewing the natural world as nothing more than a set of resources for humans to extract, therefore framing sustainable development as a way to increase the productivity of nature as a source of capital. Sustainable development was critiqued as logically inconsistent or an oxymoron, since it is inconceivable to sustain "development" infinitely in a finite world (Sauvé, Berryman, & Brunelle, 2007). Pedagogically, the term was critiqued because of the preposition "for", which is seen to promote a narrow and instrumental view of education.

More holistic views of sustainability education embrace a contemporary pedagogy of reflexivity and critical thinking in relation to contemporary discourses of sustainability (Berryman & Sauvé, 2013) and focus on meaningful actions to enhance local knowledge, democracy, and self-determination (Jickling & Wals, 2008; Wals & Jickling, 2002). The international framing of environmental education (EE), education for sustainability development (ESD) its the more pedagogically acceptable form, education for sustainability (EfS), has always been instrumental in purpose (certainly politically) for close to 50 years. It is the politics that surrounds the terminology that is problematic, not the aims, ideals and practices of the educators themselves. Any person interested in teaching and learning for a more desirable future can use international, national and state and local frameworks to support and validate their own practice.

8.3 Contemporary Australian Frameworks for Education for Sustainability

Australia is currently blessed with strong policy frameworks for environmental education and education for sustainability at national government level. In 2007, the (then) Australian Government Department of Environment and Heritage produced a document titled *Caring for our future: The Australian Government strategy for the United Nations Decade of Education for Sustainable Development* (http://aries.mq.edu.au/pdf/caring.pdf). 'Caring' was to be achieved through five principles. These were communicating the concepts of sustainability, basing the Australian Government's approach on sound research, ensuring momentum (which included developing a new national action plan), promoting a whole-of-government approach and building partnerships.

The Australian Government department responsible for environmental management is the same department responsible for meeting the UNDESD (2005–2014) international obligations. The three most important national documents framing sustainability and environmental education obligations are:

1. *Educating for a sustainable future: A National Environmental Education Statement for Australian schools* (2005) http://www.environment.gov.au/resource/educating-sustainable-future-national-environmental-education-statement-australian-schools;

2. *Living sustainability: The Australian Government's National Action Plan for education for sustainability* (2009) http://www.environment.gov.au/resource/living-sustainably-australian-governments-national-action-plan-education-sustainability
3. *Sustainability Curriculum Framework—a guide for curriculum developers and policy makers* (2010) http://www.environment.gov.au/resource/sustainability-curriculum-framework-guide-curriculum-developers-and-policy-makers.

The 2009 *National Action Plan* replaced an earlier plan published in 2000 in direct response to the UNDESD. In the Overview (pages 3 and 4), it is explained that:

Australia's approach to education for sustainability has come a long way since its origins in environmental education in the 1970s. It has evolved from a focus on awareness of natural ecosystems and their degradation to equipping all people with the knowledge, skills and understanding necessary to make decisions based upon their full environmental, social and economic implications. The diversity of sources providing education for sustainability has also grown in this time. Providers include governments, educational institutions at all levels, industry bodies, professional associations, non-government organisations, community groups, zoos, national parks, aquaria, and environmental education centres.

This last point is very important for people working in tourism enterprises. The *National Action Plan* recognizes that all organizations can provide sustainability education and that the best sustainability education is provided in partnership between different organizations, such as between a tourist organizations and education institutions. As partnerships focus on networks and relationships, the *National Action Plan* can be interpreted as strongly supporting partnerships between formal schooling and tourism and we will explore the possibilities for innovation in school-tourism partnerships later in this chapter.

The *National Action Plan* conceives a sustainable community as one that integrates environmental, social and economic goals "recognising the global dimension of sustainability; appreciating, valuing and restoring nature; conserving biodiversity and ecological integrity; ... providing for equal opportunity and community participation; committing to best practice and continuous improvement" (p. 7). Sustainability programs in schools are conceptualized around the desirability of whole school implementation (Skamp, 2010) although there is often a gap between the rhetoric of policy and the reality of implementation when it comes to environmental and sustainability education (Stevenson, 2007). The great majority of schools and early childhood education centres in Australia currently undertake sustainability education projects of some description, working toward the ideal of whole school implementation. Schools on 'the sustainability journey' are willing to investigate opportunities to engage with sustainability through partnerships with business enterprises.

Businesses interested in achieving sustainability objectives themselves are convinced to try an education for sustainability approach in an Australian Government publication titled *Taking action for the future: How organisations make successful change for sustainability* (DEWHA, 2010) (see http://www.environment.gov.au/resource/taking-action-future-how-organisations-make-successful-change-sustainability).

The core components of change within any organization are summarized as: having a vision, having a better understanding of what works (and why and when and how it works), involving all people in the organization to make sustainability changes, engaging in systemic thinking and embracing complexity to fully understand the organizations social and environmental impacts, and building strong partnerships with other organizations within industry groups and with other social sectors. A key focus in on capacity building in that:

> Capacity building using these core components increases people's ability to think differently about what needs to change and how to change it. It also improves people's confidence in making decisions that can lead to more informed action, and encourages collaborative work arrangements to achieve more effective outcomes for sustainability. (DEWHA, 2010, p. 1)

The inducement to think differently and to build people's capacity to act for positive change is also found within formal education. But how this is implemented within any state or any nation is dependent on existing state arrangements for delivery of mass education. In federated western nation such as the United States of America, Canada and Australia, state governments have responsibility for providing school education. Australia is unusual in that all state and territory governments agreed on a national curriculum that explicitly recognizes sustainability education is a function of school education. Canada does have a national framework for sustainability developed by the Council of Ministers of Education Canada (CMEC) however, there is no legal requirements for all provincial and territorial Ministries of Education to adopt the CMEC recommendations on sustainability education, though an increasing number of jurisdictions have done so (for a detailed description see Nazir, Pedretti, Wallace, Montemurro, & Inwood, 2009). In the United States, 26 (of 50) American states have agreed to the Next Generation Science Standards where students from kindergarten to Year 12 develop an understanding of socio-ecological systems thinking. These standards place emphasis on the nature of scientific evidence and students at all ages learn what evidence is; how to distinguish between weak and strong evidence; how evidence can be tested; and how "insights from many disciplines fit together into a coherent picture of the world" (NGSS 2013).

8.4 The Sustainability Cross Curriculum Priority in the Australian Curriculum

Curriculum support for whole school implementation of sustainability comes from new, national curriculum, managed by the Australian Curriculum, Assessment and Reporting Authority (ACARA, 2013). It took many years for all Australian state governments responsible for delivering school education, to agree to the creation of one national curriculum and by coincidence, the development of this curriculum took place during the UN Decade of Education for Sustainability. All was not smooth sailing when it came to placing the principles of sustainability into the national curriculum, especially within the key disciplines. Curriculum in Australia has unfortunately become highly politicized. However, after a protracted tussle as to where sustainability was going to sit within the new curriculum, the solution was to place sustainability across the disciplines (e.g. English, Mathematics, Science, History, Geography, Health and Physical Education, and the Arts).

'Sustainability' is one of three cross-curriculum priorities, along with 'Aboriginal and Torres Strait Islander histories and cultures' and 'Asia and Australia's engagement with Asia'. A cross-curriculum priority is both powerful and risky. Powerful because an across the curriculum mechanism does support whole school implementation. And risky because educators welded to the aging forms of their disciplines can ignore the priority. A cross curriculum priority can be used to leverage change, and is itself no practical impediment to change. Principals and school staff across the nation are turning their attention to how to educate their students meaningfully for the twenty-first century. We know the future will be very different from the past and that the young will bear the brunt of socio-environmental problems created over this last century, including climate change (see Nicholls & Whitehouse, 2013; Stevenson, Nicholls, & Whitehouse, 2012).

The Sustainability cross-curriculum priority is unapologetically futures focused. In the Australian Curriculum, sustainability is set out in terms echoing those of the of the 1987 *Brundtland Report* in that:

> Sustainable patterns of living meet the needs of the present without compromising the ability of future generations to meet their needs. Actions to improve sustainability are both individual and collective endeavors shared across local and global communities. They necessitate a renewed and balanced approach to the way humans interact with each other and the environment. (*Australian Curriculum* 5.2, 2013)

Sustainability is conceived around nine organizing ideas grouped into systems, world-views and futures. Organizing ideas (OIs) are a mechanism for making explicit the underlying curriculum intentions. Sustainability is focused is very much on systems thinking and developing a global view. The OIs set out the futures orientation, acknowledging we now live in conditions where, in order to get the future we want, we are going to have to make substantial changes to "the way humans interact with each other and the environment". A full elaboration of the Sustainability cross-curriculum priority can be found at: http://www.australiancurriculum.edu.au/CrossCurriculumPriorities/Sustainability.

The organising ideas (OIs) promote connection between the formal education sector and other economic sectors, such as the tourism sector especially in relation to environmental and cultural tourism, which forms a significant part of the Australian tourism market. Clear curriculum links between eco-tourism and school curriculum can be made in the Sustainability cross-curriculum priority organising ideas. For example, OI.4 states that: "World views that recognise the dependence of living things on healthy ecosystems, and value diversity and social justice are essential for achieving sustainability". OI.5 states that: "World views are formed by experiences at personal, local, national and global levels, and are linked to individual and community actions for sustainability". You can see how easily learning partnerships between schools and tourism operators can fit within the development of 'world views' within the framework of systems thinking.

Moves to embed sustainability education in the early childhood, primary school and secondary school sectors are highly congruent with moves within tertiary tourism and business education in Australia and the USA (see, for example, Benckendorff, Moscardo, & Murphy, 2012; Deale & Barber, 2012; Wilson & von der Heidt, 2013; Wilson, von den Heidt, Lamberton, & Morrison, 2012). At all formal sector levels, education for sustainability "aims to go beyond individual behaviour change and seeks to empower people to implement systemic changes" in their lives and communities (von der Heidt and Lamberton, 2011, p. 773). The first exemplar of practice (below) describes how the Great Barrier Reef Marine Park Authority (GBRMPA) acted to meet its legislative educational obligations by aligning reef education to the Australian Curriculum.

Example 1: Aligning education materials with the Australian Curriculum
Many public and private sector organizations that provide environmental education and interpretation now align their materials with the Australian Curriculum, both with the discipline areas and with the cross curriculum priorities. The reason is purely practical. For schools to make use of externally produced educational materials, these must meet the policy and curriculum settings in which teachers are required work. Well-designed educational resources and materials that are directly aligned to the national curriculum are far more likely to be implemented in schools than any other types of learning materials.

The Great Barrier Reef Marine Park Authority (GBRMPA) is an independent statutory authority with very strong partnership links to the marine tourism industry. Not only do tourism operators provide and enhance visitor experiences of costal and offshore reef systems, the tourism partners also play an important role in protecting the reef biodiversity that is the foundation of the industry. The GBRMPA is also federally legislated to provide education and environmental interpretation among its other duties. The GBRMPA encourages teachers and students across the country to undertake reef studies as well as running the extremely successful Reef Guardian stewardship program for schools, local governments and industries connected with Barrier Reef and catchments (http://www.gbrmpa.gov.au/our-partners/reef-guardians). About 300 schools are partners with the GBRMPA under this stewardship and reef "ownership" program that began in 2003.

As part of their coordinated activities to build strong links with the formal education sector, the GBRMPA developed a series of education units for Years 1–10 able to be implemented in the classroom. When the Australian Curriculum was launched, these units were rewritten to align directly with the Science discipline area of the new curriculum and with the sustainability cross-curriculum priority. The units were sent out to consultation before being made publicly available online. These science units and the additional early years activities (for very young students) and the Years 11 and 12 reef science investigation tasks are available at: http://www.gbrmpa.gov.au/resources-and-publications/student-and-teacher-resources/science-teaching-units.

The GBRMPA has also developed a series of classroom resources and online materials for use in schools across the discipline areas (not only science). These are available at: http://www.gbrmpa.gov.au/resources-and-publications/student-and-teacher-resources.

This work indicates how seriously the GBRMPA takes its educational partnership with schools along with its tourism partnerships (see http://www.gbrmpa.gov.au/our-partners/tourism-industry).

8.5 Building Partnerships for Sustainability

We only have our one small planet to live on, and the living conditions on our planet are increasingly threatened by exponential rates of development (see Rockström et al., 2009). The 'business as usual' approach to energy use and economic development is putting us in danger of abrupt climate change, high pollution and biodiversity loss and all the consequences of such. As none of us wants this, so all our work in coming decades has to be about changing our practices. Moves towards sustainability "can prepare students, teachers and [their] whole campus community to be more creative and innovative leaders in the face of complex social and biological problems, including climate change" (Wilson & von der Heidt, 2013, p. 131). The biggest challenge is, and always has been, what Wilson and von den Heidt call "operationalizing this commitment" (2013, p. 142) to sustainability education. Yes, there are many barriers, but the international and national frameworks supporting excellent practice have been in place for many years. The good news is that many people have been researching and thinking their way through these barriers. We have just concluded the United Nations Decade of Education for Sustainable Development (2005–2014). Just imagine how much work so many educators in so many nations to re-imagine a future considerate of the life systems of the planet have completed? One of the key factors for future success is partnerships and relationships.

David Orr has described sustainability as a 'recalibration'. Partnerships contribute crucially to this process by working to 'de-silo' schools and generate links with tourism enterprises that possess similar sustainability interests. If you think about it, creating partnerships is another way of doing business that acts to 'unboundary' all participating institutions—and change the resultant shape of those institutions. Benn (2010) analyses the shift to partnership as a socio-political phenomenon indicative of institutional change in post-industrial societies. Partnerships do alter the balances of power and authority associated with more traditional (i.e. hierarchical) institutions and arrangements. Partnerships are one example of what Chase-Dunn (2002) called "globalism from below".

In the twenty-first century it is useful to see partnerships between organizations, groups and individuals as *relationships for getting things done*. What needs to get done in Orr's words (2006, p. *x*) "is a process of transformation . . . in naval terms, it is time to get all hands on deck to join the fight for a habitable planet. And when the fight is joined, real learning begins". Getting "all hands on deck" is a useful metaphor for sustainability partnerships. A collaborative approach between different sectors and organizations at any level or scale can work well to promote learning provided there is a mutual, trustful relationship founded on a common purpose. The following example of practice is from a tourism interpretation enterprise in the Wet Tropics World Heritage Area.

Example 2: Private enterprise interpretation in the Daintree rainforest
The Daintree Discovery Centre was established in 1988 and is an accredited World Heritage interpretative centre located at Cow Bay over the Daintree River north of Cairns: see http://www.daintree-rec.com.au/. The environment is the collage of lowland wet forests also known as the Daintree Rainforest World Heritage area and is one of the oldest forest environments in the world, inhabited by many "dinosaur" plants—species that have existed for over 100 million years.

Concentrating on the eco-tourism experience, this interpretative organization caters for domestic and international day visitors (both individuals and tour groups) and encourages school groups. The Centre produces learning materials for both primary and secondary students and all student groups who visit the Centre have access to education officers. Interpretation is organized around the themes of experiencing the awesome ancient forest "from the forest floor to the upper most reaches of the canopy"; the collection, analysis and display of scientific data about this Australian rainforest; the threats faced by forests due to human impacts; and the role forests play in carbon offsets. (The Centre offsets its own emissions.) In addition, is the pleasurable convenience of being offered a safe and comfortable, all weather, rainforest experience. (Rainforests are not the most comfortable environment for urban humans.) Visitors move though a series of interconnected walkways; engage with layered, interactive, interpretative displays on the diversity of flora, fauna and geology; and look at how the Centre itself manages water, waste and passive cooling. School bookings are actively encouraged and guided tours and audio tours are provided.

The Centre is an example of a sustainable tourism enterprises involved in environmental interpretation and communication. According to Ham (1992), environmental interpretation is an act of translation, that is, communicating complex understandings of the natural and social world into "terms and ideas that people who are not scientists can readily understand" (p. 3). Interpretation is aimed at what Ham (1992) calls "non-captive" audiences—people who voluntarily visit a site in order to enjoy the experience of an environment and learn more about a the place they have chosen to visit. (The voluntary visitors can be school groups guided by and under the care of their teachers.) In Ham's view, the interpretative approach to communication must be pleasurable, relevant, organized and thematic—that is organized around themes rather than topics of facts. Interpretation must be engaging and entertaining in order to hold people's attention. With younger visitors, interpretation activities should be fun, well structured, meaningful for age and as hands-on as possible. Ham's guide to the technical aspects of interpretation can be found at http://classweb.gmu.edu/erodger1/prls533/Welcome/haminterp.pdf.

8.6 The Benefits and Possibilities of School-Tourism Partnerships

There are sophisticated strategies for synchronous engagement between tourism enterprises engaged with informal education and interpretation, and the formal education sector. Partnerships for sustainability have been promoted internationally since the 1987 *Bruntland Report* as they maximize the benefits of different expertise, capacities, and resources available. Partnerships are synergistic, not just one-way transfers of services or knowledge. Timpson et al. (2006, p. 10) put it thus: "We live in a system; we work in a system; our economy, government and environment are all systems. This realization can be overwhelming and confusing: it can make anyone feel inadequate and unable to influence change." Because any system has so many parts, the key to making change happen across a system is forming learning partnerships across different parts (sectors) of our systems.

Partnership relationships coordinate sustainability action across different sectors and act *as leverage* to effect systemic change. Learning is essential to sustainability partnerships when the task is to overcome the inertia of 'business as usual' thinking.

Tourism within any geography is created from a myriad of cooperative practices, "collaborative arrangements" and "cross-sectoral interactions" (Bramwell & Lane, 2000). In assessing the potential for tourism partnerships to contribute to the wider social objectives of promoting sustainability, Bramwell and Lane (2000, p. 4) write that "collaboration among a range of stakeholders, including non-economic interests [such as educational interests] might promote more consideration of the varied natural, built and human resources that need to be sustained for present and future well-being". One of the payoffs for tourism organizations to partner with local schools is the impact factor. Australian Bureau of Statistics data show that 19 % of Australians attend school either as students or staff. Even if we assume an influence factor of one school attendee to one other community member (1:1), the community impact factor is a minimum 40 %. In reality, the impact factor is usually greater that 50 %, and even higher is smaller communities. This means that the all the positive messages generated by a successful partnership will travel extensively through the community.

Successful partnerships are successful relationships that can also be formalized with partnership agreements and memoranda of understanding. Successful partnerships save duplication of economic, skills, cultural and knowledge resources. The other dimensions are also rewarding, such as the social dimension of collaboration;

the new insights, approaches, and capabilities that emerge; and the growth of the intelligence of participating organizations.

Recently, RMIT published the results of an ARC funded research project into school—community partnerships for sustainability as a guidebook for practitioners. The monograph is called *Conversations on School-Community Learning Partnerships for Sustainability: A Guidebook* (Smith, Wheeler, Guevara, Gough, & Fien, 2012). The full guidebook can be downloaded at: http://mams.rmit.edu.au/s3ysio6sumic1.pdf. This project showed that locale is very important. Educational institutions serving children and adolescents are geographically located within local communities that are also served by a whole range businesses including tourism businesses, local level government (often very useful partners for schools), state, and federal government departments; non-government and scientific organizations. Partnerships enable young people and their teachers to access key expertise (people who know a lot about energy, waste, water, gardening and biodiversity, climate change); additional resources (direct cash grants or in-kind support); and direct access to learning experiences, expert environmental interpretation and local environmental knowledge. Research showed that strong partnerships between schools and local businesses and community organizations have real impact on developing socio-ecological awareness and knowledge leading to a willingness to take local action (Smith et al., 2012).

The third example of practice concerns a very successful Cairns tourism enterprise and its focus on forming learning partnerships for cultural education.

Example 3: A sustainable partnership for cultural learning

Tjabukai Aboriginal Cultural Park (http://www.tjabukai.com.au/) located at Caravonica, Cairns, is privately owned by Indigenous Business Australia. Its vision is to "restore, the culture, customs and traditions of the Tjabukai people" and its mission "is to educate and entertain our visitors". Over three million people have experienced the Tjabukai Aboriginal Cultural Park.

Education marketing is one of the organization's key interests. Going to 'Tjabukai' is great fun and highly informative and the Park has won numerous tourism awards. A day or night trip to the Park is an activity supported by Education Queensland (Queensland state Department of Education, Training and Employment). There are many elements to this reciprocity. Schools who bring 15 students or more are provided with an education officer. Park employees have written student workbooks from Year 1 to Year 12 and the bookshop carries a range of relevant education materials. Twice a year, teachers can attend a free promotional day to learn about the Park, and, the Park provides teachers with a downloadable School Assessment Risk Assessment Guide for their excursion. Every school that visits the Park is issued an admission voucher to be used for school fundraising within the following 12 months. Tjabukai Aboriginal Cultural Park is a "flagship" of Australian cultural tourism, bringing the knowledge and stories of the Bama Balumba (rainforest people) to the world. The Park also works very closely with the local community and with the education sector, from early childhood to university level. Every story told, every dance enhances people's understanding of the 40,000 year-old aboriginal rainforest culture. 'Tjabukai' is changing our contemporary culture though wit, humour and gentle education.

Wheatley (2001, p. 67), points out that "an organization rich with many interpretations develops a wiser sense of what is going on and what needs to be done. Such organizations become more intelligent." Wheatley argues to "go for the jazz" (or the dance) and get involved together and "improvise". While change can only ripple though a system with persistent effort, small changes that pay instant rewards to the partnership are often the best way to progress. When people report on their short-term successes, this creates a positive feedback loop that strengthens the partnership to do more over a longer time. Successful partnerships can really un-stick old practices. Leadership that seeks to develop trust and collaboration in order to bring about innovations and improvements in student experience will be a key factor in any successful school-tourism enterprise partnerships. Tjabukai Aboriginal Cultural Park is a highly trusted organization, which is why its partnership arrangement with Education Queensland is one of long-standing.

8.7 Conclusion

The future we will inherit is the one in which we are in the process of creating. It is important to understand that schools are not silos and schools will be located at various points along a change continuum in terms of educating for a more sustainable future. As the Decade of Education for Sustainable Development (UNDESD, 2005–2014) comes to an end, many conclusions have been reached as to the possibilities and practicalities of sustainability education. For one, education for sustainability "promotes learning beyond the boundaries of educational institutions" to equip people "with the knowledge, skills and values to address [the] social, environmental and economic challenges of the 21st century" including preparing for jobs that "preserve or restore the quality of the environment" (UNESCO 2013). For another, education for sustainability promotes all manner of generative partnerships. The idea is to think past hierarchical barriers and look sideways instead to create networks and relationships for action and learning. We have looked at three exemplars of successful partnerships between the education and tourism sectors in far north Queensland. There are many variations of this theme of cooperation and accommodation. The formal education sector is porous. The tourism enterprise sector has a huge impact to make in terms of educating and interpreting for sustainability. There are many opportunities to go and explore.

References

Australian Curriculum Assessment and Reporting Authority. (2013). Australian Curriculum 5.2. http://www.australiancurriculum.edu.au. Accessed 20 Nov 2013.

Benckendorff, P., Moscardo, G., & Murphy, L. (2012). Environmental attitudes of Gen Y students: Foundations for sustainability education in tourism. *Journal of Teaching in Travel and Tourism, 12*(1), 44–69.

Benn, S. (2010). *Social partnerships for governance and learning towards sustainability*. Sydney: Australian Research Institute for Environment and Sustainability, Macquarie University.

Berryman, T., & Sauvé, L. (2013). Languages and discourses of education, environment and sustainable development. In R. B. Stevenson, M. Brody, J. Dillon, & A. E. J. Walls (Eds.), *The international handbook of research on environmental education*. New York: Routledge.

Blaze Corcoran, P. (2010). Sustainability education in higher education: Perspectives and practices across the curriculum. In P. Jones, D. Selby, & S. Stirling (Eds.), *Sustainability education. Perspectives and practice across higher education* (pp. x–xii). New York: Earthscan.

Bramwell, B., & Lane, B. (2000). Collaboration and partnerships in tourism planning. In B. Bramwell & B. Lane (Eds.), *Tourism collaboration and partnerships: Politics, practice and sustainability* (pp. 1–19). Clevedon: Channel View.

Chase-Dunn, C. (2002). Globalisation from below: Toward a collectively rational and democratic global commonwealth. *Annals of the American Academy of Political and Social Science, 581*, 48–61.

Deale, C. S., & Barber, N. (2012). How important is sustainability education to hospitality programs? *Journal of Teaching in Travel and Tourism, 12*(2), 165–187.

Department of Environment and Heritage. (2010). *Taking action for the future: How organisations make successful change for sustainability*. Canberra: Australian Government Department of Water, Heritage and the Arts.

Gough, A. (1997). *Education and the environment: Policy, trends and the problem of marginalisation*. Melbourne: Australian Council for Education Research.

Ham, S. H. (1992). *Environmental interpretation: A practical guide for people with big ideas and small budgets*. Golden: North American Press.

Jickling, B., & Wals, A. E. J. (2008). Globalization and environmental education: Looking beyond sustainable development. *Journal of Curriculum Studies, 40*(1), 1–21.

Nazir, J., Pedretti, E., Wallace, J., Montemurro, D., & Inwood, H. (2009). Climate change and sustainable development: A response from education in Canada. Ontario Institute for Studies in Education (Research report to the IALEI). http://www.hilaryinwood.ca/research/climate_change_sustainable_development.html#SomeFactsaboutCanada. Accessed 10 Jan 2014.

Next Generation Science Standards for Today's Students and Tomorrow's Workforce. (2013). http://www.nextgenscience.org. Accessed 10 Jan 2014.

Nicholls, J., & Whitehouse, H. (2013). Climate change education in the primary and middle years of schooling. *Primary and Middle Years Educator, 11*(2), 10–15.

Orr, D. (2006). All hands on deck—Teaching sustainability. In W. M. Timpson, B. Dunbar, G. Kimmel, B. Bruyere, P. Newman, & H. Mizia (Eds.), *147 tips for teaching sustainability: Connecting the environment, the economy and society* (pp. ix–x). Madison, WI: Arwood.

Poole, R. (2010). *Earthrise: How man first saw the earth*. New Haven, CT: Yale University Press.

Rockström, J., Steffen, W., Noone, K., Perrson, Å., Chapin, F. S., 3rd, Lambin, E. F., et al. (2009). A safe operating space for humanity. *Nature, 461*, 472–475.

Sauvé, L., Berryman, T., & Brunelle, R. (2007). Three decades of international guidelines for environment-related education: A critical hermeneutic of the United Nations discourse. *Canadian Journal of Environmental Education, 12*, 33–54.

Skamp, K. (2010). *Critical review of current practice and research of environmental education and education for sustainability for Kindergarten to Year 10 from 1990*. Sydney: New South Wales Department of Education and Training.

Smith, J.-A., Wheeler, L., Guevara, J. R., Gough, A., & Fien, J. (2012). *Conversations on school community learning partnerships for sustainability (a guidebook)*. Bundoora: RMIT University.

Stevenson, R. B. (2007). Schooling and environmental education: Contradictions in purpose and practice. *Environmental Education Research, 13*(2), 139–153.

Stevenson, R. B., Nicholls, J., & Whitehouse, H. (2012). Challenges for educators of building people's capacity for mitigating and adapting to climate change. In A. E. J. Wals & P. Blaze Corcoran (Eds.), *Learning for sustainability in times of accelerating change* (pp. 365–380). Amsterdam: Wageningen.

Timpson, W. M., Dunbar, B., Kimmel, G., Bruyere, B., Newman, P., & Mizia, H. (2006). *147 practical tips for teaching sustainability*. Madison, WI: Arwood.

UNESCO World Conference on ESD 2014. (2013). Five reasons to support ESD—Education for sustainable development. http://www.unesco.org/new/index.php?id=96295. Accessed 24 Nov 2013.

Von der Heidt, T., & Lamberton, G. (2011). Sustainability in the undergraduate and postgraduate business curriculum of a regional university: A critical perspective. *Journal of Management and Organisation, 17*(5), 672–692.

Walls, A. E. J., & Jickling, B. (2002). "Sustainability" in higher education: From doublethink and newspeak to critical thinking and meaningful learning. *International Journal of Sustainability in Higher Education, 3*(3), 221–232.

Walls, A. E. J., & Jickling, B. (2008). Globalization and environmental education: Looking beyond sustainable development. *Journal of Curriculum Studies, 40*(1), 1–21.

Wheatley, M. (2001). *Leadership and the new science*. San Francisco: Barret-Koehler.

Wilson, E., & von der Heidt, T. (2013). Business as usual? Barriers to education for sustainability in the tourism curriculum. *Journal of Teaching and Travel in Tourism, 13*(2), 130–147.

Wilson, E., von den Heidt, T., Lamberton, G., & Morrison, D. (2012, June). *Are we moving towards education for sustainability? A study of sustainability embeddedness in a first year undergraduate business/tourism curriculum*. Presentation to BEST-EN Sustainable Tourism Think Tank XII, Greoux les Bairns.

Part III
Education for Sustainability in Tourism: Key Stakeholders

Chapter 9
Educating Destination Communities for Sustainability in Tourism

Gianna Moscardo and Laurie Murphy

Abstract The people who live and work in destination communities are crucial, but often neglected, stakeholders in tourism. The literature on community involvement in tourism management and research into tourism impacts on destinations indicates that the involvement of destination communities in tourism governance is a key element of sustainability. This chapter will focus on residents of destination communities and outline the educational elements required to support more effective community based governance of tourism. It will also explore the ways in which tourism governance experiences could be used to support Education for Sustainability (EfS) in general and community action for sustainability beyond tourism.

Keywords Community involvement • Stakeholder • Public participation • Sustainable tourism • Governance

9.1 Introduction

Tourism has a long been used as a tool for community development based on the assumption that it creates economic opportunities for local businesses and residents and generates financial capital that can then be used to generate other benefits for the destination community (Moscardo, 2008a, 2008b). There is, however, little evidence that supports this assumption with many studies reporting extensive negative impacts from tourism in a number of domains (Moscardo, 2008a; Murphy, 2013). These negative impacts on the destination environment and community are one key reason why tourism has long been linked to sustainability issues (Saarinen, 2013). Education for Sustainability (EfS) in destination communities therefore needs to address elements required to support better planning for the development and management of tourism to mitigate negative impacts on destination community well-being (DCW). A key factor in this tourism sustainability challenge is the need for effective citizen participation in tourism governance (Bramwell & Lane, 2011). This participation in tourism governance also offers an opportunity to use the

G. Moscardo (✉) • L. Murphy
James Cook University, Townsville, QLD, Australia
e-mail: Gianna.moscardo@jcu.edu.au; Laurie.Murphy@jcu.edu.au

G. Moscardo, P. Benckendorff (eds.), *Education for Sustainability in Tourism*, CSR, Sustainability, Ethics & Governance, DOI 10.1007/978-3-662-47470-9_9

experience of tourism to support EfS more generally within destination communities. A similar duality of purpose is noted for EfS with tourists in Chap. 11 of this book on Sustainability Education for Tourists.

In order to address these two issues this chapter will firstly examine the relationships between tourism and destination community well-being (DCW), briefly reviewing research into how tourism contributes to, and/or detracts from, the various types of capital that are said to make up DCW. This brief review will highlight the importance of effective community involvement in tourism planning and management and the critical role that tourism knowledge plays in supporting this. The chapter will then suggest what would be required from education to support more effective community involvement in tourism and a positive role for tourism in sustainability at the destination level. The discussion of EfS for destination communities will particularly focus on the importance of understanding community well-being requirements and the use of tourism as a tool for building sustainability capacity within destination communities. Examples from a current action research project on tourism, sustainability and DCW will be used to highlight these themes.

Before beginning this examination of EfS for tourism with destination communities, it is important to consider what is meant by community in the present discussion. Lyon and Driskell (2012) offer a detailed history and critical analysis of the concept of community in the social sciences. They describe a number of different approaches including community as a spatial entity, community as a social network, community as a common lifestyle, and community as a shared set of goals, values and practices. They go on to offer a definition of community as "*people in a specific area who share common ties and interact with one another*" (Lyon & Driskell, 2012, p. 5) and argue that this definition fits the most common uses of the concept. It is also the definition that best aligns with discussions of tourist destinations. Thus a destination community is defined for this chapter as the people who live and work within the spatial boundaries of a tourist destination. The authors recognise that there are multiple types of residence and work possible within any community, and that this definition includes the tourism businesses that operate within the destination. However, as tourism businesses and their staff have been examined in another chapter within this volume, the main focus of this chapter will be on residents living in and around destinations.

9.2 Linking Tourism to Sustainability Through Destination Community Well-Being

The concepts of quality of life (QoL) and well-being are closely linked to sustainability, with sustainability concerns driving growing public interest in measuring and managing aspects of life beyond economic growth and financial wealth (Aspinall, Cukier, & Doberstein, 2011; UNDP, 2013). This is discussed further in Chap. 2. A number of models of sustainability incorporate QoL and wellbeing as core elements arguing that a wide range of factors need to be considered in moving to more sustainable practices (Costanza et al., 2007; Lehtonen, 2004; Vermuri & Costanza, 2006). While QoL and well-being are often used interchangeably, for the purposes of clarity in this chapter QoL will be used when discussing individuals and well-being will be used for the social and community level. Both concepts are based on the underlying argument that positive outcomes depend upon the degree to which an individual or community is able to meet a set of basic needs (Barwais, 2011). Despite multiple classification systems and diverse terminology, there is considerable consensus about what these needs are and they include:

- Physiological needs such as drinkable water, food, health, and physical protection;
- Security needs such as a safe and stable place to live and work;
- Belongingness needs including access to social networks, and opportunities to be part of social, cultural and political activities; and
- Self-esteem needs such as the ability to have the confidence and knowledge to make decisions and opportunities to engage in valued activities (Clarke, Islam, & Paech, 2006; Costanza et al., 2007; Malkina-Pykh & Pykh, 2008).

These well-being needs are met by access to different types of capital, which are described in Table 9.1. This argument that destination community well-being (DCW) is significantly influenced by tourism impacts on these different types of capital is one that is being given increasing attention in the tourism literature (Andereck & Nyaupane, 2010; Macbeth, Carson, & Northcote, 2004; McGehee, Lee, O'Bannon, & Perdue, 2010; Moscardo, 2009, 2012; Moscardo, Konovalov, Murphy, & McGehee, 2013). This approach to understanding tourism impacts on

Table 9.1 Types of capital necessary to support destination community well-being

Capital	Characteristics
Financial	Monetary assets, savings and access to these assets for investment in production or for exchange into other capitals
Physical	Infrastructure available to support other activities including buildings, transport, energy, resources and telecommunication systems
Natural	The systems, processes and resources that are provided in the natural environment and that support biological life and human activities
Human	The skills, abilities, health, knowledge and availability of individuals that can be used to support production and other activities
Social	The relationships, group memberships, connections and networks, trust and good will that exists within groups and allows access to other capitals and social support
Cultural	The store of shared knowledge and material elements needed to support cultural expression through art, craft and ritual
Political	The mechanisms and capacity to access and influence relevant political decision-making processes

Based on Emory and Flora (2006) and Lehtonen (2004)

these different dimensions of DCW offers a way to both critically analyse tourism processes and to improve the sustainability of tourism.

Of particular importance is the need to understand how different features of tourism and its development and management contribute to longer-term sustainable development in destination communities. It is possible to describe the positive and negative impacts that tourism can have on each of these types of capital. Studies by Nunkoo and colleagues (Nunkoo and Ramkissoon, 2012; Nunkoo and Smith, 2013), for example, describe the different ways tourism can impact on one element of social capital, destination residents' trust in government institutions. Similar research can be found in the areas of natural capital (cf. Hernández & León, 2007), human capital (Zapata, Hall, Lindo, & Vanderschaeghe, 2011), and cultural capital (Snowball and Willis, 2006).

While these tourism impacts on the various types of capital that make up DCW provide some insights into ways to improve tourism outcomes for destinations, discussions of tourism and sustainability must also consider the wider and cumulative impacts of tourism across the whole system of interactions between these different forms of capital. Reviews of case studies describing tourism development in a range of different destination communities provide a way to do this. Such reviews consistently identify two key conclusions. The first is that overall tourism is rarely an effective development option and there is little evidence that it makes an overall or net positive contribution to the well-being of destination communities (Hall, 2011; Moscardo, 2008a, 2014; Ruhanen, 2004). The second is the identification of barriers to effective and sustainable tourism development outcomes, connected in a consistent pattern as presented in Fig. 9.1. This figure shows a sequence of actions that cumulatively contribute to poor outcomes for tourism as a community development option. A common thread through these factors is a lack of knowledge and understanding of tourism and how it operates as a system and a lack of skills related to effective tourism governance, especially amongst those most likely to be making tourism development decisions, which results in limited citizen involvement in tourism planning or management and through that ineffective and unsustainable tourism (Moscardo, 2011a).

Citizen involvement, also known as public or community participation, in governance and local empowerment in development decision-making is seen as a core but challenging element of sustainability (Laessoe, 2010; Mackelworth & Carić, 2010). Most discussions of participatory governance of sustainability are based on the concept of a ladder of participation as proposed by Arnstein (1969) and adapted by Brager and Specht (1973) and Pretty (1995). Table 9.2 provides a summary of the main elements of these various ladders of participation. The different systems share a similar progression from a total lack of community involvement in development decisions and actions in the lowest row, through situations where the community is given information about what it is happening and then asked to comment on the proposed developments but without any control or power to alter or reject the proposals, to the higher levels where the community has greater levels of power and control, finishing with the community initiating and managing the development process in its entirety.

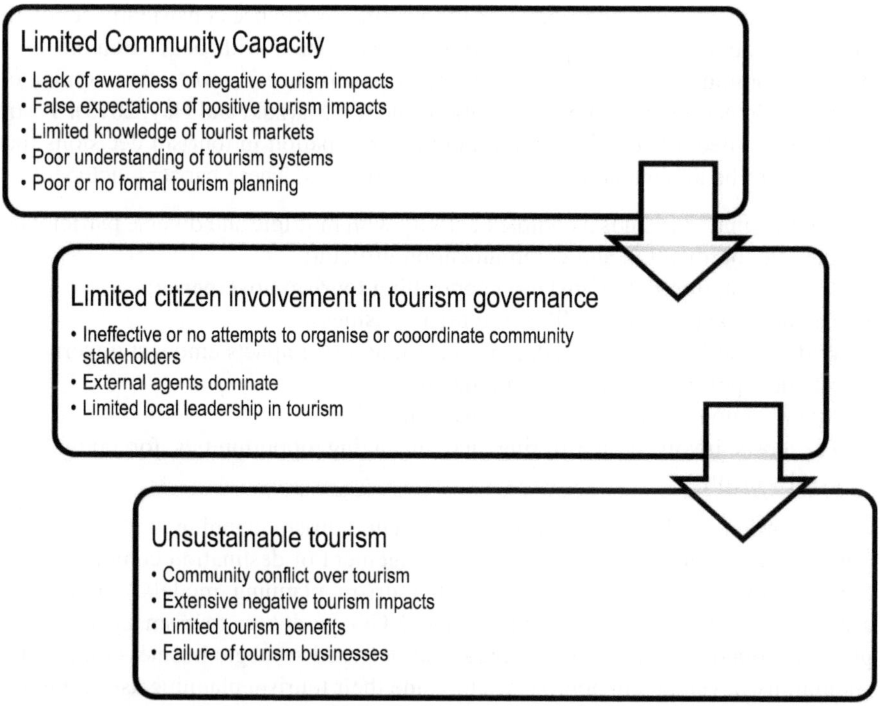

Fig. 9.1 Barriers to effective and sustainable tourism development (based on Moscardo, 2011a)

Table 9.2 Ladders of public participation in governance

Arnstein (1969)	Brager and Specht (1973)	Pretty (1995)	Marzuki and Hay (2013)
• Citizen control • Community delegates power • Community developer partnership	• Community has control • Community delegates authority • Community & developer plan jointly	• Self-mobilisation • Incentive participation	• Empowerment
• Placation • Consultation	• Community advises • Community is consulted	• Functional participation • Consultation	• Consultation
• Information	• Community given information	• Passive participation	• Information
• Therapy • Manipulation	• No community involvement	• Manipulative participation	

Discussions of public participation in tourism governance consistently conclude that while higher levels of participation are associated with better outcomes, in practice community involvement in tourism governance is most often conducted within the lower levels of the participation ladders (Marzuki and Hay, 2013). Tosun (2000) identified a list of barriers to public participation in tourism decisions that continues to be applicable in contemporary situations. These barriers included:

- The fact that tourism is a complex activity with multiple and diverse participants making coordination and communication difficult;
- A lack of appropriate legislation controlling tourism governance;
- The dominance of local elites in tourism business;
- Widespread lack of knowledge of tourism and its impacts amongst government officials, politicians, NGOs and citizens;
- Apathy amongst destination citizens; and
- The costs involved in offering and managing opportunities for community involvement.

Moscardo's (2011b) analysis of the dominant tourism planning models suggests a further set of barriers to the effective engagement of destination communities in tourism governance. This review of 36 tourism planning models dominant in academic textbooks and government and NGO guides to tourism development planning found that most tourism organisations used strategic business rather than community development approaches to guide their tourism planning assuming that tourism was desirable and necessary and that the primary objective of tourism planning was to encourage tourists to visit, not to support DCW. Destination communities were rarely included as key stakeholders or given significant roles in decision-making and the major sustainability issues recognised were environmental. This analysis suggests that the fundamental assumptions made in traditional tourism planning processes exclude and disempower destination communities and see destinations as resources for tourism, rather than tourism as a tool or resource for destinations.

This brief review of the relationships between destination communities, their well-being, and tourism suggests that EfS in tourism requires attention to two areas:

- Alternative approaches to tourism planning processes and governance; and
- Education of destination community stakeholders about the nature of tourism as a global activity, its impacts and their management which supports the development of skills and capacity for community participation in tourism governance.

9.3 Alternative Approaches to Tourism Governance: A Case Study

The authors have been involved in an ongoing action research project based on the use of tourism as a tool to support the sustainability objectives of a tourist destination community which includes the development of a DCW approach to tourism planning (see Moscardo & Murphy, 2013 for more details). This project offers a case study that highlights some of the features that would be required in new approaches to tourism governance and provides insights into the education required to support such a new approach.

Magnetic Island (MI) is located 8 km off the north eastern coast of Australia adjacent to the major regional city of Townsville and within the boundaries of the Great Barrier Reef World Heritage Area (Dept. of National Parks, Recreation, Sport & Racing, 2011). Sixty-five percent of its 52 km^2 area is a National Park and it is officially a suburb of Townsville and home to approximately 2300 residents (Aust. Bureau of Statistics, 2010). MI has been a popular recreational spot since European settlement of the region in the 1800s with a slow increase over an extended period of time of small scale, locally owned tourist businesses.

In 1984 a large scale, tourist resort complex was proposed and this triggered more than 20 years of community conflict both amongst MI residents and between MI residents and a succession of tourism development companies (Moscardo & Murphy, 2013). This conflict and the resulting negative environmental, social and economic impacts on this community have been described in detail elsewhere (Harrington, 2004; Heywood, 1990; Valentine, 1989) but are noteworthy in the present context for four main reasons. Firstly, a critical element contributing to the conflict was the lack of effective public participation processes in the decisions made about this tourism development proposal (Heywood, 1990). Secondly, the current community is generally suspicious of tourism development proposals whilst recognising the potential importance of tourism as a necessary economic activity in the destination (Moyle, Croy, & Weiler, 2010). Thirdly, the community has developed a strong focus on sustainable living evident in a activities such as a successful solar energy program (Bruce, Heslop, Macgill, & Watt, 2013) and significant reductions in water usage (Citiwater Townsville, 2002). Finally, there exist a number of MI community based organisations that developed out of the tourism development conflicts that support community action and public participation in governance in general.

Figure 9.2 provides an overview of the proposed new approach which conceptualizes tourism as tool to support DCW aspirations and needs. This approach differs from the traditional approaches described in Moscardo (2011b) in three main ways because it:

- Places the destination community at the centre of the tourism planning process with input into every stage rather than as a stakeholder to be informed after key decisions have been made;

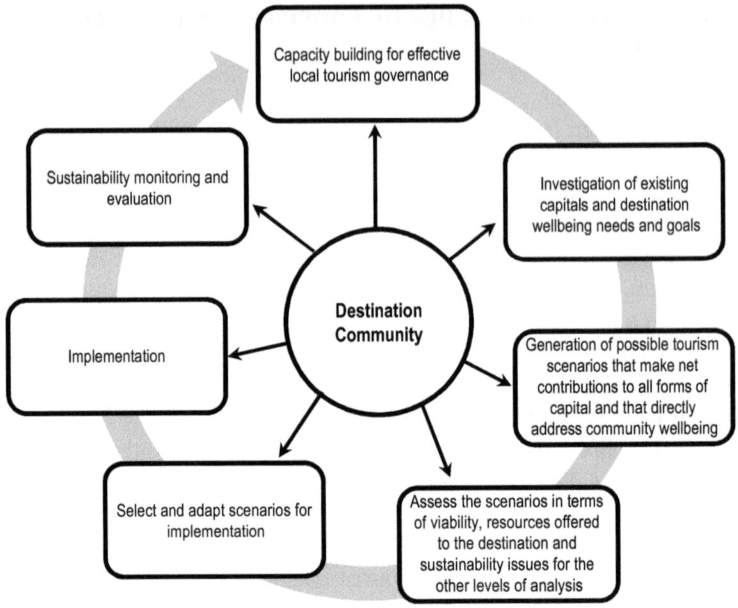

Fig. 9.2 A community well-being approach to destination tourism planning (Moscardo & Murphy, 2013, p. 2547)

- Begins by suggesting that communities need to build capacity for tourism governance;
- Suggests that tourism needs to be considered specifically within the context of DCW needs and aspiration; and
- Requires the tourism planning process to explicitly involve future scenario development and assessment.

The MI research project offered an opportunity to apply parts of this new approach to a specific destination. Of particular importance to the present chapter were the outcomes of resident workshops using futures wheel and backcasting techniques (see Benckendorff, 2008; Benckendorff et al., 2009 for more details on these approaches) to focus attention on the ways in which tourism could directly contribute to key aspects of DCW. The first step in the workshop was to inform participants of the key results of research into the profile of current visitors and non-visitors and their perceptions of the island and of potential sustainable tourism product development initiatives. This research was conducted in response to stakeholder concerns and was designed with input from a range of stakeholders. The workshop then sought to focus participant attention on the key dimensions of, and improvements to, DCW. For this destination a strong sense of community, which was linked to small scale development, friendliness and personal involvement in community activities, and being closely connected to the natural environment were the most important dimensions of well-being. DCW needs were identified in four main areas:

Fig. 9.3 Summary futures wheel linking tourism to MI DCW

- Better public transport;
- Maintenance and enhancement of health and education services;
- Maintaining, supporting and extending various sustainability initiatives; and
- A desire to alter the style of tourism to be more consistent with, and directly supportive of, the community's sustainable lifestyle aspirations.

This last area of DCW aspirations supported a more detailed examination of how tourism might be able to support improvements to MI DCW and Fig. 9.3 includes the summary futures wheel generated from these activities.

The futures wheels exercise revealed that efforts to attract the types of visitors who are more likely to be sympathetic to the lifestyle of island residents, to educate these visitors on environmental issues such as MI's solar and sustainability initiatives, and to improve interpretation of local marine and land based environments, are important concerns for the locals. The participants identified a need for better understanding of tourist markets, improved facilities and resources for information and education provision, and a focus on attracting styles and scales of development which are smaller, greener and within an identified carrying capacity for the island, as necessary to achieve their tourism aspirations. A desire for the island to maintain and develop its own sense of place, distinctive from other tropical resort-based islands along the coast was also emphasised. Importantly more effective communication structures within the community and between tourism operators were also highlighted. This is particularly relevant given the multiple community interest groups that exist on the island, challenges in attracting membership to the tourism business group, and difficulty in achieving consensus and a shared vision for tourism.

The project has contributed to a shift in tourism priorities towards educational tourism strategies that are more compatible with community aspirations. These are different to the priorities identified in previous cycles of more traditional tourism development planning that proposed strategies for MI that included the building of 4–5 star resort properties that would provide more than 100 additional rooms, and the inclusion of MI in a cluster of Great Barrier Reef islands for promotional campaigns focussed on adventure, coastal and cruise tourism (Tourism Queensland, 2007).

There was a clear message from the participants that while the potential contribution of tourism to the local economy is important, tourism development should not come at the cost of negative impacts on either the quality of life of locals or the island's natural environment. While there is evidence in the responses of a recognition of the potential for tourism to contribute to DCW beyond generating income for locals, there is also an indication of knowledge gaps and skill development issues which need to be addressed, with a particular focus on providing and incorporating insight on tourist markets into the destination marketing and distribution strategies and collaborative approaches to governing and planning for tourism incorporating the existing fragmented and often contested views of sub-groups within the local community.

9.4 Educating Destination Community Stakeholders About Tourism: Themes, Skills, and Methods

Discussions of education in the tourism development literature have most often been about training staff to work in tourism and/or training local business people to become tourism entrepreneurs (Cooper, Fletcher, Fyall, Gilbert, & Wanhill, 2008; WTO, 2007). The aim of education in this context is to support the production of tourism opportunities and to meet the perceived needs of tour operators and tourists and typically includes information on tourism and hospitality management practice, service skills and some attention to understanding markets and tourist expectations (cf. Shakeela, Breakey, & Ruhanen, 2012). While it can be argued that these skills are necessary to support local resident participation in tourism business, this approach still assumes that the goal of tourism is to create financial capital with little critical reflection on whether or not this contributes to an overall increase in DCW.

The MI workshops revealed that destination communities want a better understanding of tourism and tourists so that they can make more informed decisions about the types of tourists and tourist businesses they want to attract and support. This is consistent with the literature on public participation in general and in tourism specifically and suggests a different set of topics or themes in education to support more effective and higher level community involvement in tourism governance. Education in this context needs to be *about* tourism rather than *for* tourism (Cole, 2005; Haywood, 1988; Marzuki and Hay, 2013; Moscardo, 2011a; Potts, 2010; Timothy and Tosun, 2003). Discussions of the education needed to support effective community engagement in sustainability governance in general suggest three main areas that need attention—knowledge/information about the relevant activity (in this case tourism), skills to support participation in governance and methods or opportunities for learning and engagement (Sarkissian, Hofer, Shore, Vajda, & Wilkinson, 2008; Tilbury and Wortman, 2004).

In terms of knowledge about tourism, the core requirement is to understand the nature of tourism and the systems involved in its production, distribution and consumption (Cole, 2005; Dredge and Moore, 1992; Gartner, 1996; Moscardo, 2011a). More specifically within this tourism system, destination communities need to have knowledge in five key areas. The first is understanding the range, extent and nature of both the positive and potentially negative impacts of tourism including knowledge about the processes and factors that contribute to these impacts (Chakravarty, 2003; Keogh, 1990; Moscardo, 2011a; Upchurch and Teivane, 2000). Secondly, these impacts need to be considered in the context of information about the range of different types of tourism development. It is particularly important to be able to assess and compare the feasibility, costs and benefits and financial risks associated with different types of tourism (Chakravarty, 2003; Keogh, 1990). The third area is an understanding of the nature of tourism consumption including the characteristics and motivations of tourists, tourist decision-making processes and managing tourist behaviour (Moscardo, 2011a). The fourth area concerns the competitive nature of tourism, the ways in which tourism is distributed and managed, the requirements for its production in destination regions and the connections between tourism and other activities in the region (Moscardo, 2008b; Potts, 2010). Finally, EfS in tourism needs to include an examination of the opportunities for local businesses and residents to participate in, and benefit from, tourism (Chakravarty, 2003; Keogh, 1990).

In addition to a more detailed and critical understanding of tourism as a system and its impacts, destination communities also need to have a series of skills for using that knowledge and effectively engaging in tourism governance. There is considerable consensus about what these skills are within the tourism and broader sustainability literature and they include many of those listed for EfS in general (see Chap. 1) such as:

- Networking and partnership development and management skills;
- Collaboration and group facilitation skills, including conflict resolution and negotiation skills;
- Knowledge management and dissemination skills;
- Cross-cultural and diversity skills;
- Social entrepreneurship,
- Goal setting, visioning and future scenario building skills;
- Change management; and
- Responding to uncertainty and complexity (Okazaki, 2008; Powell, Bailey, Wade, & Tunstall, 2014; Roseland, 2012; Sarkissian et al., 2008; Temenos and McCann, 2012; Tilbury and Wortman, 2004; Wiseman, Williamson, & Fritze, 2010).

Although there is recognition of the need to educate communities about tourism with consensus about the topics to include in this education, there has been very little research or description of methods for such community education programs. There are a few exceptions to this, with some studies of specific educational techniques reported for emerging destinations in peripheral regions where residents often have little experience of being tourists, limited access to formal education and lower levels of literacy. Cole (2006a, 2006b), for example, provides a description and evaluation of a community education program conducted in Indonesia which used community workshops, focus groups and interaction with a tourism researcher through an action research program. Timothy (2000), in another Indonesian case study, described a program that included vocational training, short courses and a mass media campaign supported by brochures. Sammy (2008) described the use of drama and drawings to support tourism information exchange between destination residents in several African locations and NGO staff. Sammy (2008) also supported the use of field trips in which key residents from one potential destination community travel to other places with more established tourism ventures to see them in operation and talk about the challenges and lessons learnt. Stronza (2008) describes a similar process with additional 3–5 day structured workshops conducted in South America. In these examples the authors stress the importance of facilitators having a long-term relationship with the target communities. The Panos Institute's (nd) review of campaigns conducted in parts of the Caribbean to raise public awareness of tourism's potential negative environmental impacts advocated the use of community events, news media, websites that provide access to resources, seminars and workshops.

Although there have been few systematic evaluations of community education programs about tourism, it is possible to identify techniques and principles from the more common programs conducted to encourage destination residents to support tourism and to work for tourism. Techniques that have been used for these types of tourism awareness campaigns include:

- Television and/or radio advertising;
- Articles, regular columns and/or programs on various news media;
- Public/community events;

- Newsletters or brochures sent directly to residents or through schools, community and church groups, shopping precincts and government offices;
- Posters;
- Websites that provide access to resource kits and programs (cf. Australian Department of Environment and Heritage, 2004; Mountain Institute, 2000 for examples of these types of resources);
- Public meetings;
- Seminars and workshops;
- Short courses;
- Taking residents on commercial tours or visits to existing tourist facilities;
- Resource packages for school teachers;
- Activities and curriculum materials for primary and secondary school students;
- Essay, poster and photography competitions; and
- Events with policy or decision-makers.

(see Caribbean Tourism Organization, nd; Empower-Ed, 2004; Indian Institute of Tourism and Travel Management, 2011; USAID, 2009 for examples of these types of programs). The evaluations of the programs that have been conducted indicate that effectiveness is more likely when:

- Techniques are matched to the audience in terms of their most common communication channels and levels of existing knowledge;
- A variety of techniques are used;
- Information is widely accessible; and
- Programs are in an appropriate language (Indian Institute of Tourism & Travel Management, 2011; Panos Institute, nd).

These suggestions are consistent with those made for public education campaigns in general and the theories that underpin effective persuasive communication. Recent edited books by van Trijp (2014) and Rice and Atkin (2013) provide insights into these areas.

Education alone is not sufficient to improve sustainability, there also needs to be changes to the opportunities for citizen involvement in tourism governance. As with the previous discussion on the abilities required for improved community participation in tourism and sustainability governance, there is considerable convergence in the literature about the changes needed to public participation methods and these are summarised in Table 9.3. These changes require commitment both from community members and those in power. Government organisations need to be prepared to give up at least some of their authority and control and citizens need to be prepared to take on responsibility (Hung, Sirakaya-Turk, & Ingram, 2011; Mackelworth & Carić, 2010; Marzuki and Hay, 2013; Roseland, 2012; Tilbury, 2004).

According to Hung et al. (2011) effective public participation in tourism governance requires skills and abilities, opportunity and motivation. To date the motivation or willingness on the part of citizens to get involved in tourism governance has been acknowledged as an issue by many authors, with very few offering specific suggestions for addressing this challenge (cf. Hung et al., 2011; Marzuki

Table 9.3 Principles for improving opportunities for public participation

Suggested principles
• More extensive programs in terms of both the length of time allowed for public participation and the use of a diverse range of methods including surveys, interviews, public meetings/ community forums, handbooks, public education campaigns through local media, and toolkits
• Use of newer media and computer technologies to present information and offer opportunities for input including the use of websites, blogs, wikis (cf. Farina et al., 2011 for a discussion of the use of social media, Brabham, 2012 for a discussion of crowdsourcing; and Wang and Bryer (2013) for an evaluation of an online discussion forum and blog)
• Using existing local networks, opinion leaders and gate keepers to establish groups and partnerships to support the public participation
• Gathering and disseminating relevant knowledge including sustainability audits and monitoring
• Creating formal co-management structures and organisations
• Using multiple languages and media
• Paying particular attention to accessing and involving vulnerable and hard to reach groups

Sources: Mackelworth and Carić (2010); Roseland (2012); Sarkissian et al. (2008); Tilbury and Wortman (2004); Wiseman et al. (2010)

and Hay, 2013). Research into programs with higher levels of empowerment, where the public has real control and is listened to, and that use a wider range of public participation opportunities suggests that citizen apathy may be a response to the nature of the traditional processes and that it is possible to generate positive public attitudes towards participation (Brabham, 2012; Byrd, 2007; Conrad, Cassar, Christie, & Fazey, 2011; Evely, Pinard, Reed, & Fazey, 2011).

9.5 Tourism Governance as a Tool for EfS

A recurring theme in the literature on EfS for community participation and engagement in sustainability decisions and actions is that of the importance and value of action-oriented learning or learning by doing (see Chap. 1). Resident participation in tourism governance and/or community based sustainable tourism projects offer an opportunity for this kind of learning (Laessoe, 2010; Temenos & McCann, 2012; Tosun, 2000). Tourism has a number of features that make it particularly useful for learning to support sustainability more generally. It is multi-sectoral and requires linkages between many different types of organisations, between the public and private sectors and between the destination community and external groups, and it relies on facilities and services that can be used by residents as well as visitors and so it can be used to generate tangible benefits linked to participation (Moscardo, 2012).

Potts (2010) provides an example of how a community activity focussed on improving the sustainability of tourism to the Blue Mountains in Australia encouraged a focus on environmental protection and supported the creation of positive business partnerships that in turn encouraged further sustainability action within the community. Similarly, Hamilton and Alexander (2013) describe an example of a community tourism initiative in Scotland that provided residents with tangible benefits in terms of access to rejuvenated public and social space and highlighted their shared cultural heritage. Matarrita-Cascante, Brennan, and Luloff (2010) working in Costa Rica provide evidence that the construction of local relationships for a community based tourism project increased community capacity to respond to other sustainability challenges, highlighted the value of local knowledge and made residents aware of their shared values and the importance of stewardship of local cultural and environmental resources. All these outcomes of programs with high levels of local resident empowerment in tourism decisions demonstrate how tourism might be used as a social learning tool for sustainability in general.

9.6 Conclusions and Challenges

This chapter has discussed the dimensions of, and suggested principles for, EfS for tourism with destination community residents. This discussion has argued if EfS is to be effective in this context there needs to be change in tourism development and governance practice. In the case of tourism development this requires a move towards more explicit consideration of destination community well-being and in the case of tourism governance there has to be a move towards greater empowerment for destination residents. This also has implications for changes in tourism within the higher education sector.

The EfS literature in general consistently argues that experiential learning opportunities that connect students to communities are a critical element of sustainability curricula (Cotton & Winter, 2010; Wade, 2013) suggesting the need for more action research projects where tourism and other students, along with educators and researchers work with local communities to enhance tourism sustainability. Such programs could particularly focus on the role of tourism groups within higher education as providers of the types of credible and reliable information about tourism and its impacts that Wiseman et al. (2010) have argued is necessary for effective community sustainability action. In turn this emphasizes the importance of teaching students change management and leadership skills (Rowe & Johnston, 2013; Tilbury, 2013). It also challenges higher education providers to see alternative career pathways for tourism graduates including roles as community change agents and social entrepreneurs (Roseland, 2012).

Working on tourism sustainability projects with communities also offers an opportunity for tourism students and researchers to better understand the local cultural, historical and political contexts for community sustainability action. It has been suggested that it may not always be appropriate to expect wide ranging democratic public participation in sustainability and that assumptions about the importance of citizen involvement in sustainability action are based on western worldviews that may not apply everywhere (Wesselink, Paavola, Fritsch, & Renn, 2011). Examples from practice in places such as China (cf. Boland & Zhu, 2012) and Croatia (cf. Mackelworth & Carić, 2010) suggest, however, that it is possible to adapt governance to local conditions and learning by doing encourages this awareness and capacity.

References

Andereck, K. L., & Nyaupane, G. P. (2010). Exploring the nature of tourism and quality of life perceptions among residents. *Journal of Travel Research, 50*(3), 248–260.

Arnstein, S. R. (1969). A ladder of citizen participation. *Journal of the American Institute of planners, 35*(4), 216–224.

Aspinall, A., Cukier, J., & Doberstein, B. (2011). Quality of life assessments and social sustainability: Ski tourism development in Invermere, British Columbia. *Journal Environmental Assessment Policy and Management, 13*(2), 179–201.

Australian Bureau of Statistics. (2010). *National regional profile: Magnetic Island.* Available Online: http://www.abs.gov.au/ausstats/abs@nrp.nsf/781eb7868cee03e9ca2571800082bece/2a3e7368b29b7557ca2577d50011bdaa!OpenDocument. Accessed 11 Nov 2013.

Australian Department of Environment and Heritage. (2004). *Steps to sustainable tourism.* http://www.environment.gov.au/system/files/resources/9c8f5084-3565-4f26-8801-77da3458f497/files/steps.pdf. Accessed 5 Feb 2015.

Barwais, F. (2011). *Definitions of wellbeing, Quality of life and wellness.* National Wellness Institute of Australia. http://nwia.idwelness.org/2011/02/28/definitions-of-wellbeing-quality-of-life-and-wellness/. Accessed 12 Feb 2104.

Benckendorff, P. (2008). Envisioning sustainable tourism futures: An evaluation of the futures wheel method. *Tourism and Hospitality Research, 8*(1), 25–36.

Benckendorff, P., Edwards, D., Jurowski, C., Liburd, J. J., Miller, G., & Moscardo, G. (2009). Exploring the future of tourism and quality of life. *Tourism and Hospitality Research, 9*(2), 171–183.

Boland, A., & Zhu, J. (2012). Public participation in China's green communities: Mobilizing memories and structuring incentives. *Geoforum, 43*(1), 147–157.

Brabham, D. C. (2012). The effectiveness of crowdsourcing public participation in a planning context. *First Monday, 17*(12). http://journals.uic.edu/ojs/index.php/fm/article/view/4225. Accessed 11 Nov 2013.

Brager, G., & Specht, H. (1973). *Community organizing.* New York: Columbia University Press.

Bramwell, B., & Lane, B. (2011). Critical research on the governance of tourism and sustainability. *Journal of Sustainable Tourism, 19*(4-5), 411–421.

Bruce, A., Heslop, S., Macgill, I., & Watt, M. (2013). *Magnetic Island and Townsville Solar City.* Canberra: Australian Renewable Energy Agency.

Byrd, E. T. (2007). Stakeholders in sustainable tourism development and their roles: Applying stakeholder theory to sustainable tourism development. *Tourism Review, 62*(2), 6–13.

Caribbean Tourism Organization. (n.d). Competing with the best: Good practices in tourism awareness programmes. http://www.onecaribbean.org/content/files/GoodPracticestourismawareness%282%29.pdf. Accessed 5 Feb 2015.

Chakravarty, I. (2003). Marine ecotourism and regional development: A case study of the proposed Marine Park at Malvan, Maharashtra, India. In B. Garrod & J. Wilson (Eds.), *Marine ecotourism: Issues and experiences* (pp. 177–197). Clevedon: Channelview.

Citiwater Townsville. (2002). *Magnetic Island water recycling.* Townsville: Citiwater.

Clarke, M., Islam, S. M. N., & Paech, S. (2006). Measuring Australia's well-being using hierarchical needs. *Journal of Socio-Economics, 35*, 933–945.

Cole, S. (2005). Community education. In D. Airey & J. Tribe (Eds.), *An international handbook of tourism education* (pp. 395–407). Abingdon: Routledge.

Cole, S. (2006a). Cultural tourism, community participation and empowerment. In M. Smith & M. Robinson (Eds.), *Cultural tourism in a changing world: Politics, participation and (re) presentation* (pp. 89–103). Clevedon: Channelview.

Cole, S. (2006b). Information and empowerment: The keys to achieving sustainable tourism. *Journal of Sustainable Tourism, 14*(6), 629–644.

Conrad, E., Cassar, L. F., Christie, M., & Fazey, I. (2011). Hearing but not listening? A participatory assessment of public participation in planning. *Environment and Planning-Part C, 29* (5), 761.

Cooper, C., Fletcher, J., Fyall, A., Gilbert, D., & Wanhill, S. (2008). *Tourism: Principles and practice* (4th ed.). Essex: Pearson.

Costanza, R., Fisher, B., Ali, S., Beer, C., Bond, L., Boumans, R., et al. (2007). Quality of life: An approach integrating opportunities, human needs, and subjective well-being. *Ecological Economics, 61*, 267–276.

Cotton, D., & Winter, J. (2010). 'It's not just bits of paper and light bulbs': A review of sustainability pedagogies and their potential for use in higher education. In P. Jones, D. Selby, & S. Sterling (Eds.), *Sustainability education: Perspectives and practice across higher education* (pp. 39–54). Abingdon: Earthscan.

Department of National Parks, Recreation, Sport and Racing. (2011). *About Magnetic Island.* Available online: http://www.nprsr.qld.gov.au/parks/magnetic-island/about.html. Accessed 11 Nov 2013.

Dredge, D., & Moore, S. (1992). A methodology for the integration of tourism in town planning. *Journal of Tourism Studies, 3*(1), 8–21.

Emory, M., & Flora, C. B. (2006). Spiraling-up: Mapping community transformation with community capitals framework. *Community Development, 37*, 19–35.

Empower-Ed. (2004). *Buffalo City municipality tourism master plan: Phase 9, 10 and 12*. http://www.buffalocity.gov.za/municipality/keydocs/tourism_masterplan/phase10_12/phase9_10_12.pdf. Accessed 5 Feb 2015.

Evely, A. C., Pinard, M., Reed, M. S., & Fazey, I. (2011). High levels of participation in conservation projects enhance learning. *Conservation Letters, 4*(2), 116–126.

Farina, C. R., Miller, P., Newhart, M. J., Cardie, C., Cosley, D., & Vernon, R. (2011). Rulemaking in 140 characters or less: Social networking and public participation in rulemaking. *Pace Law Review, 31*, 382–463.

Gartner, W. C. (1996). *Tourism development: Principles, processes, and policies*. New York: Van Nostrand Reinhold.

Hall, C. M. (2011). Policy learning and policy failure in sustainable tourism governance: From first-and second-order to third-order change? *Journal of Sustainable Tourism, 19*(4-5), 649–671.

Hamilton, K., & Alexander, M. (2013). Organic community tourism: A cocreated approach. *Annals of Tourism Research, 42*, 169–190.

Harrington, J. T. (2004). *Being here: Heritage, belonging and place making*. PhD Thesis, James Cook University, Townsville.

Haywood, K. M. (1988). Responsible and responsive tourism planning in the community. *Tourism Management, 9*(2), 105–118.

Hernández, J. M., & León, C. J. (2007). The interactions between natural and physical capitals in the tourist lifecycle model. *Ecological Economics, 62*(1), 184–193.

Heywood, P. (1990). Social justice and planning for the public interest. *Urban Policy and Research, 8*(2), 60–68.

Hung, K., Sirakaya-Turk, E., & Ingram, L. J. (2011). Testing the efficacy of an integrative model for community participation. *Journal of Travel Research, 50*(3), 276–288.

Indian Institute of Tourism and Travel Management. (2011). *Evaluation of social awareness campaign for good behaviour towards tourists*. http://tourism.gov.in/writereaddata/CMSPagePicture/file/marketresearch/studyreports/SocialAwarenessReport.pdf. Accessed 5 Feb 2015.
Keogh, B. (1990). Public participation in community tourism planning. *Annals of Tourism Research, 17*(3), 449–465.
Laessoe, T. (2010). Education for sustainable development, participation and socio-cultural change. *Environmental Education Research, 6*(1), 39–57.
Lehtonen, M. (2004). The environmental-social interface of sustainable development: Capabilities, social capital, institutions. *Ecological Economics, 49*, 199–214.
Lyon, L., & Driskell, R. (2012). *The community in urban society* (2nd ed.). Long Grove: Waveland Press.
Macbeth, J., Carson, K., & Northcote, J. (2004). Social capital, tourism and regional development: SPCC as a basis for innovation and sustainability. *Current Issues in Tourism, 7*(6), 502–522.
Mackelworth, P. C., & Carić, H. (2010). Gatekeepers of island communities: Exploring the pillars of sustainable development. *Environment, Development and Sustainability, 12*(4), 463–480.
Malkina-Pykh, I. G., & Pykh, Y. A. (2008). Quality-of-life indicators at different scales: Theoretical background. *Ecological Indicators, 8*, 854–862.
Marzuki, A., & Hay, I. (2013). Towards a public participation framework in tourism planning. *Tourism Planning & Development, 10*(4), 494–512.
Matarrita-Cascante, D., Brennan, M. A., & Luloff, A. E. (2010). Community agency and sustainable tourism development: The case of La Fortuna, Costa Rica. *Journal of Sustainable Tourism, 18*(6), 735–756.
McGehee, N. G., Lee, S., O'Bannon, T., & Perdue, R. (2010). Tourism-related social capital and its relationship with other forms of capital: An exploratory study. *Journal of Travel Research, 49*, 486–500.
Moscardo, G. (2008a). Community capacity building—An emerging challenge for tourism development. In G. Moscardo (Ed.), *Building community capacity for tourism* (pp. 1–15). Wallingford: CABI.
Moscardo, G. (2008b). Sustainable tourism innovation: Challenging basic assumptions. *Tourism and Hospitality Research, 8*(1), 4–13.
Moscardo, G. (2009). Tourism and quality of life: Towards a more critical approach. *Tourism and Hospitality Research, 9*, 159–170.
Moscardo, G. (2011a). The role of knowledge in good governance for tourism. In E. Laws, H. Richins, J. Agrusa, & N. Scott (Eds.), *Tourist destination governance* (pp. 67–80). Wallingford: CABI.
Moscardo, G. (2011b). Exploring social representations of tourism planning: Issues for governance. *Journal of Sustainable Tourism, 19*(4-5), 423–436.
Moscardo, G. (2012). Building social capital to enhance the quality of life of destination residents. In M. Uysal, R. Perdue, & M. J. Sirgy (Eds.), *Handbook of tourism and quality of life research* (pp. 403–422). New York: Springer.
Moscardo, G. (2014). Tourism and community leadership in rural regions: Linking mobility, entrepreneurship, tourism development and community well-being. *Tourism Planning & Development, 111*(3), 1–17.
Moscardo, G., Konovalov, E., Murphy, L., & McGehee, N. (2013). Mobilities, community wellbeing and sustainable tourism. *Journal of Sustainable Tourism, 21*(4), 532–556.
Moscardo, G., & Murphy, L. (2013). There's no such thing as sustainable tourism: Reconceptualizing tourism as a tool for sustainability. *Sustainability, 6*(1), 2538–2561.
Mountain Institute. (2000). *Community based tourism for conservation and development: A resource kit*. http://mountain.org/sites/default/files/attachments/community_based_tourism_for_conservation_and_development.pdf. Accessed 5 Feb 2015.
Moyle, B. B., Croy, G., & Weiler, B. (2010). Community perceptions of tourism: Bruny and Magnetic Islands, Australia. *Asia Pacific Journal of Tourism Research, 15*(3), 353–366.

Murphy, P. E. (2013). *Tourism: A community approach* (Vol. 4). London: Routledge.
Nunkoo, R., & Ramkissoon, H. (2012). Power, trust, social exchange and community support. *Annals of Tourism Research, 39*(2), 997–1023.
Nunkoo, R., & Smith, S. L. (2013). Political economy of tourism: Trust in government actors, political support, and their determinants. *Tourism Management, 36*, 120–132.
Okazaki, E. (2008). A community-based tourism model: Its conception and use. *Journal of Sustainable Tourism, 16*(5), 511–529.
Panos Institute. (n.d.). *Improving training and public awareness on Caribbean coastal tourism.* http://www.cep.unep.org/issues/panos.PDF. Accessed 5 Feb 2015.
Potts, T. (2010). The natural advantage of regions: Linking sustainability, innovation, and regional development in Australia. *Journal of Cleaner Production, 18*(8), 713–725.
Powell, R., Bailey, R., Wade, A., & Tunstall, J. (2014). Reaching out to local communities in a discussion of sustainable development. *Bridges,* 10. doi:9780784412688-021.
Pretty, J. N. (1995). Participatory learning for sustainable agriculture. *World Development, 23*(8), 1247–1263.
Rice, R. E., & Atkin, C. K. (2013). *Public communication campaigns* (4th ed.). Los Angeles: Sage.
Roseland, M. (2012). *Toward sustainable communities: Solutions for citizens and their governments*. Vancouver: New Society.
Rowe, D., & Johnston, L. F. (2013). Learning outcomes: An international comparison of countries and declarations. In L. F. Johnston (Ed.), *Higher education for sustainability* (pp. 45–59). New York: Routledge.
Ruhanen, L. (2004). Strategic planning for local tourism destinations: An analysis of tourism plans. *Tourism and Hospitality Planning & Development, 1*(3), 239–253.
Saarinen, J. (2013). Critical sustainability: Setting the limits to growth and responsibility in tourism. *Sustainability, 6*(1), 1–17.
Sammy, J. (2008). Examples of effective techniques for enhancing community understanding of tourism. In G. Moscardo (Ed.), *Building community capacity for tourism development* (pp. 75–85). Wallingford: CABI.
Sarkissian, W., Hofer, N., Shore, Y., Vajda, S., & Wilkinson, C. (2008). *Kitchen table sustainability*. London: Earthscan.
Shakeela, A., Breakey, N., & Ruhanen, L. (2012). Tourism education's roles in sustainable tourism development: A case study of SIDS introduction. *Journal of Hospitality & Tourism Education, 24*(1), 35–43.
Snowball, J. D., & Willis, K. G. (2006). Building cultural capital: Transforming the South African national arts festival. *South African Journal of Economics, 74*(1), 20–33.
Stronza, A. (2008). Partnerships for tourism development. In G. Moscardo (Ed.), *Building community capacity for tourism development* (pp. 101–115). Wallingford: CABI.
Temenos, C., & McCann, E. (2012). The local politics of policy mobility: Learning, persuasion, and the production of a municipal sustainability fix. *Environment and Planning-Part A, 44*(6), 1389–1406.
Tilbury, D. (2004). Rising to the challenge: Education for sustainability in Australia. *Australian Journal of Environmental Education, 20*(2), 103–114.
Tilbury, D. (2013). Another world is desirable: A global rebooting of higher education for sustainable development. In S. Sterling, L. Maxey, & H. Luna (Eds.), *The sustainable university* (pp. 71–85). Abingdon: Routledge.
Tilbury, D., & Wortman, D. (2004). *Engaging people in sustainability*. Gland: IUCN.
Timothy, D. J. (2000). Building community awareness of tourism in a developing country destination. *Tourism Recreation Research, 25*(2), 111–116.
Timothy, D. J., & Tosun, C. (2003). Appropriate planning for tourism in destination communities: Participation, incremental growth and collaboration. In S. Singh, D. J. Timothy, & R. K. Dowling (Eds.), *Tourism in destination communities* (pp. 181–204). Wallingford: CABI.
Tosun, C. (2000). Limits to community participation in the tourism development process in developing countries. *Tourism Management, 21*(6), 613–633.

Tourism Queensland. (2007). *Destination management plan for tourism in Townsville, North Queensland, 2007-2010*. Brisbane: Tourism Queensland.

UNDP. (2013). *Human development report. The rise of the South: Human progress in a diverse world.* http://www.undp.org/content/dam/philippines/docs/HDR/HDR2013%20Report%20English.pdf. Accessed 12 Feb 2014.

Upchurch, R. S., & Teivane, U. (2000). Resident perceptions of tourism development in Riga, Latvia. *Tourism Management, 21*(5), 499–507.

USAID. (2009). Tourism awareness plan 2010-2011. http://pdf.usaid.gov/pdf_docs/PNADR524.pdf. Accessed 5 Feb 2015.

Valentine, P. (1989). Community response to tourism proposals on Magnetic Island. *Urban Policy and Research, 7*(2), 83–85.

van Trijp, H. C. M. (2014). *Encouraging sustainable behavior: Psychology and the environment*. New York: Psychology Press.

Vermuri, A. W., & Costanza, R. (2006). The role of human, social, built and natural capital in explaining life satisfaction at the country level: Toward a national well-being index. *Ecological Economics, 58*, 119–133.

Wade, R. (2013). Promoting sustainable communities locally and globally. In S. Sterling, L. Maxey, & H. Luna (Eds.), *The sustainable university* (pp. 89–151). Abingdon: Routledge.

Wang, X., & Bryer, T. A. (2013). Assessing the costs of public participation: A case study of two online participation mechanisms. *American Review of Public Administration, 43*(2), 179–199.

Wesselink, A., Paavola, J., Fritsch, O., & Renn, O. (2011). Rationales for public participation in environmental policy and governance: Practitioners' perspectives. *Environment and Planning-Part A, 43*(11), 2688.

Wiseman, J., Williamson, L., & Fritze, J. (2010). Community engagement and climate change: Learning from recent Australian experience. *International Journal of Climate Change Strategies and Management, 2*(2), 134–147.

WTO. (2007). *A practical guide to destination management*. Madrid: WTO.

Zapata, M. J., Hall, C. M., Lindo, P., & Vanderschaeghe, M. (2011). Can community-based tourism contribute to development and poverty alleviation? Lessons from Nicaragua. *Current Issues in Tourism, 14*(8), 725–749.

Chapter 10
Education for Sustainability in Tourism: Coaching Tourism Businesses

Anna Blackman and Sharee Bauld

Abstract It is often espoused that micro and small enterprises (MSEs) are the engines of growth of a country's economy. There are, however, considerable barriers that they face in start up and operations that are compounded by their inability to utilise the necessary resources to assist them. As such more targeted information delivered directly to MSE owner/managers, supported by flexible, personalised programs for capacity building and person support in combination with peer networks, could provide a more effective and sustainable approach to encourage MSEs to incorporate sustainable tourism development responses into their business practices. This chapter will explore the practice of business coaching as an education tool and participatory learning method for tourism MSEs adopting a sustainable development approach to their business practices. In particular, how business coaching can best integrate economic, environmental, and social concepts into core tourism business activities is examined. The chapter will conclude with a practical guide on how to implement coaching programmes for tourism MSEs.

Keywords Coaching • Tourism • Sustainability • Education • Business

10.1 Introduction

Tourism is often proposed as an alternative strategy for businesses, in regions where traditional industries such as agriculture and manufacturing are experiencing an economic downturn. In promoting tourism development there has often been an approach of 'build it and the tourists will come', without practical consideration for the challenges involved in developing and promoting demand driven, sustainable tourism products or services. Tourism is often cited as the new agent of change for destinations; one that maintains the integrity and quality of life of the location's setting, producing new economic opportunities that traditional industries once brought (Garrod, Warnell, & Youell, 2006; Lane, 1994).

A. Blackman (✉) • S. Bauld
James Cook University, Townsville, QLD, Australia
e-mail: anna.blackman@jcu.edu.au; sharee@bauld.com

G. Moscardo, P. Benckendorff (eds.), *Education for Sustainability in Tourism*, CSR, Sustainability, Ethics & Governance, DOI 10.1007/978-3-662-47470-9_10

It is often espoused that MSEs are the engines of growth of a country's economy (Gebremichael, 2014; Khan, 2013; Nuwagaba & Nzewi, 2013; Ongoro, Kiragu, & Simwa, 2013; Sharma & Gounder, 2011). There are, however, considerable barriers that they face in start up and operations which are compounded by their inability to utilise the necessary resources to assist them (Biggs, Hall, & Stoeckl, 2012). As such more targeted information delivered directly to MSE owner/managers, supported by flexible, personalised programs for capacity building and person support in combination with peer networks, could provide a more effective and sustainable approach to encourage MSEs to incorporate sustainable tourism development responses into their business practices.

The integration of sustainability into business practice requires a paradigm shift in thinking, culture and everyday businesses practice (Tilbury, 2004) which can deter MSE owners/managers from adopting or considering sustainable practices as part of their core business model. Businesses big and small struggle with the increasing pressure of integrating sustainability into business practice. While this may be due to a level of complacency of the business owner/manager, it may also be the case that the knowledge and skills required to effect change is unknown and difficult to acquire (Tilbury, 2004).

To examine the potential for tourism to build sustainable practices, it is first important to consider tourism as a system which stresses the interconnectedness between the demand- (the market) and supply-side (transportation, attractions, services and promotion), as well as with external elements such as the natural environment, cultural resources, social structures including organisation and leadership, community attitudes, availability of finance and entrepreneurs, competition and government policies (Gunn, 1988). Seeing tourism as a comprehensive system enables recognition of the web of linkages between the industry and the broader community (Bauer, Sofield, Webb, Battig, & De Lacy, nd) and enables consideration of new avenues for development of business opportunities beyond those that are traditionally considered 'tourism'. However in destinations where the shift to tourism is occurring, many businesses have still not linked in with the industry. Studies suggest the need for capacity building in and around tourism to better understand the challenges of entering new industries and how to overcome these (Blackman et al., 2004).

This chapter will explore the practice of business coaching as an education tool and participatory learning method for tourism MSEs adopting a sustainable development approach to their business practices. In particular, how business coaching can best integrate economic, environmental, and social concepts into core tourism business activities is examined. The chapter will conclude with a practical guide on how to implement coaching programmes for tourism MSEs.

10.2 Sustainable Tourism Destinations

In areas where tourism is seen as a viable economic option, a sustainable tourism approach is needed to safeguard the integrity of these areas, conserving the local environment and social norms and improving on the quality of life. Tourism's contribution towards job creation and economic development is often seen as the linchpin to the successful development of regions which in turn helps to sustain and improve the quality of life in these areas (Lordkipanidze, Brezet, & Backman, 2005; Sharpley, 2001). However, the development of tourism needs to include a broader acceptance by the community, of development that meets the needs of the local people (Bramwell, 1994). This type of approach lends itself to the concept of developing social capital, whereby stakeholders are able to make more informed decisions about the degree to which a community adopts tourism (Macbeth, Carson, & Northcote, 2004). According to Macbeth et al. (2004: 505) "social capital is the bank of resources built up through interpersonal networks and associations upon which individual members of a community can draw". These resources (networks, associations, reciprocity, civic participation and connections) when enacted, work to contribute to the innovative capacity of destinations and regions to develop and maintain a viable tourism industry, requiring the coming together of Regional Tourism Organisations, local governments, industry associations and MSEs to collectively create new initiatives, jointly advocate on issues, seek advice and discuss options, and cooperate in partnerships, particularly in areas where access to services is limited (Macbeth et al., 2004). The act of stakeholders networking for a common cause creates a situation of leadership that is also conducive to more collaborative destinations. The coming together of stakeholders in this type of setting can help to create a stronger regional tourism system, particularly where there is a need to integrate a complex set of interconnected interests. The need for networking to create stronger linkages between destinations and their surrounding regions is critical if tourism is to be used successfully as an agent for regional development. If communities contain the necessary social capital with which to build and maintain a sustainable tourism industry, integrated regional areas will have a better chance of survival when external impacts occur (Macbeth et al., 2004).

A lack of access to information for MSEs, means it is particularly important for stakeholders to understand the importance of, and links between, tourism and sustainable development. While key sustainable development issues need to be incorporated such as biodiversity, sustainable consumption, etc. equally important are the need to adopt sustainable business practices to ensure the longevity of the industry. Furthermore, there is the case for needing to highlight tourism as an example of how sustainable development thinking can be applied to addressing destination development problems. With tourism education and sustainability the mainstay of the higher education realm, it is seldom taught to destination communities, tourism businesses, or other key tourism stakeholders (Moscardo, 2008). As such, an absence of education for these stakeholders limits their capacity to effectively engage in tourism decision making at the destination level, and is often the reason for ensuing unsustainable tourism development (Moscardo, 2011).

10.3 Sustainable Tourism MSEs

Micro (defined as 1–4 employees) and small (defined as 5–19 employees) enterprises (MSEs) play an important role in the economy providing employment and a wide range of goods and services in Australia (Gibb, 2009; Murta, Gero, Kuruppu, & Mukheibir, 2012). They are most likely to be found in agriculture, fishing, construction, tourism and retail (Murta et al., 2012). In Australia, New South Wales (33 %) and Victoria (25 %) were the states with the largest number of registered MSEs, followed by Queensland (20 %). MSEs employ 47 % of the Australian population of which 85 % is within the services sector. Of this, tourism (accommodation and food services) accounts for 53 % or just under half a million people. MSEs contribute 35 % of industry value-add. Over 95 % of the total MSE contribution to industry value added is attributable to the services sector, of which 43 % is considered tourism (accommodation and food services) or $13.4 billion to the Australian economy. As of 2010, MSEs comprised 96 % of all registered businesses of which micro-businesses accounted for 24 % of total businesses and small business (11 %) (non-employing business was 60 %) (Commonwealth of Australia, 2011).

In the development of tourism destinations it is important for key stakeholders to understand the importance of, and links between, tourism and sustainable development. One way to provide this understanding is through education. The UN for Education for Sustainable Development states that education for sustainable development "requires participatory teaching and learning methods that motivate and empower learners to change their behaviour and take action for sustainable development," where traditional teaching methods are replaced by new practices that best assist in this new learning environment. These new learning methods are said to encourage the development of certain competencies such as critical thinking, envisioning future scenarios and collaborative decision making (UNESCO, nd). In considering these new learning methods, a range of management techniques and practices are available to support the capacity building of MSEs in areas such as sustainable development including mentoring, job assignment and action learning. Each of these tools has been developed within the context of a structured organisation and so can be difficult to apply and manage in a smaller and less formally structured business such as an MSE. However, two options which are more suited to building capacity within this context are classroom programs (e.g. workshops) (Day, 2001), and business coaching (Day, 2001; Hall, Otazo, & Hollenback, 1999; Peterson, 1996). Business coaching is a helpful management technique for MSEs as it develops their individual goals for the business with one-on-one coaching to provide a continual process of goal redevelopment and achievement.

For an MSE to embrace sustainability both internally and externally, an important aspect of the learning process is to enable the integration of sustainable development principles into core business practices, rather than simply learning about sustainable development principles. Traditional forms of education have centred on the practice of learning about sustainability, which has focused on understanding the key sustainability issues affecting businesses such as climate change, resource scarcity and ethical sourcing. However learning or education for sustainability (EfS) goes beyond traditional education methods to incorporate a more holistic approach to learning, that utilises the application of a wider set of skills necessary to respond to sustainability issues, whilst embedding knowledge into everyday business practice (see Chap. 1 of this book). As sustainability needs to be considered from a "holistic, critical and futures-oriented" perspective, so too does the vehicle within which it is delivered. Fundamental and long lasting shifts in thinking, practice and culture, is a requirement of sustainability and something that can be difficult to implement in businesses—large or small (Tilbury, 2004: 109).

With MSEs being the dominant contributor to the tourism economy, they often face many challenges in remaining commercially viable. Adding to these complexities is the challenge of these small owner-operated businesses in applying sustainable development principles to their business practices (Carlsen, Getz, & Ali-Knight, 2001). Importantly therefore MSEs need to be part of the sustainable development process in order to remain viable. Tourism MSEs need to adopt the triple-bottom-line approach not only for their own businesses but for the sector as a whole.

10.3.1 The Triple Bottom Line for MSEs

Sustainability for MSEs is about achieving sustainable operations and the triple bottom line approach i.e. the integration of economic, environmental, and social concepts into core business activities. The practicality of sustainable tourism means adopting a 'best practice approach', suggesting the integration of corrective or improved measures into a businesses management and operations with the goal of ensuring that "the least possible impact is caused, that tourist product quality and image are improved, that business development becomes more efficient, and therefore, social and economic development does as well" (Rainforest Alliance, 2005: 5).

MSEs are often viewed as one of the vehicles by which the economic, social and cultural benefits of tourism can be distributed. Roberts and Tribe (2008) argue that the sheer number of tourism MSEs and their increasing role within the framework of contributing to the sustainable development of destinations, suggests that MSEs have the potential to help these destinations progress in their objective to adopt approaches to develop sustainable tourism destinations. This means MSEs are beginning to play an ever increasing and key role in sustainable tourism development.

When examining the triple bottom line approach to sustainability, the three concepts are discussed briefly. At its most basic level, economic sustainability refers to a business's ability to make a profit in order to survive and benefit the local economic system (Roberts & Tribe, 2008). Economic sustainability is about building linkages and reducing leakages; essentially the notion that tourism profits are retained within the local community. It's about generating more prosperous communities through tourism activity. In the context of MSEs, businesses that are economically efficient help to distribute the benefits more widely throughout the local community, employing others and buying local goods and services.

In integrating environmental concepts into core business activities, Goodall and Stabler (1992) suggest sustainability must be examined from both a destination and a business operation point of view. Suggestions for adopting various environmental concepts into a business included "developing environmental friendly tourism products, complying with environmental regulations, reducing any negative impacts of business operations, increasing the efficiency of resource use, using (and substituting other supplies for) environmentally friendly products, fostering an understanding of environmental issues (e.g. among visitors) and implementing codes of conduct" (Goodall & Stabler, 1992: 19). Denman (1994: 217) also provided criteria for the integration of environmental concepts into sustainable tourism development, including "sensitive development appropriate to the local environment (e.g. using traditional buildings/building styles), support for conservation projects and the economy at a very local level (i.e. retain profits locally), ecologically sound practices, and providing visitors with a genuine appreciation of the area".

Social sustainability for an MSE means giving back to the community, the community benefiting from the distribution of profits from tourism, and reducing the impacts communities may experience from tourism development. These measures can have a "positive effect on employees, and this benefits both the employee and the business" creating a more conducive working environment, increased productivity, and improved businesses reputation, which ultimately provides a competitive advantage for the business and the destination as a whole (Barton, 2010).

In the discussion on sustainability and MSEs, the aim is to make businesses reassess their operations in favour of development that is not just economically sound, but also socially/culturally, and environmentally. Achieving the triple bottom line is often prescribed for business sustainability. However, implementing sustainability further requires businesses to change their behaviour and current practices in order to take action on issues of sustainability. The challenge for MSEs lies in acquiring management habits and tools that will support the business as it develops economically, socially and environmentally. This is where a tool such as coaching can be of benefit.

10.4 Coaching Tourism MSEs for Sustainability

As previously mentioned, EfS goes beyond traditional education methods. Learning beyond the bounds of the traditional education system is said to include a number of dimensions. One of these refers to the need to address life-long learning. Life-long learning, beyond the formal classroom, is "seen as an approach to everyday living based on a predisposition to think about and reflect critically on one's experiences and be willing to change" (Moscardo, Chap. 1: 5). In this sense, the practice of business coaching can be said to better suit the approach of life-long learning, addressing the requirements of MSEs to shift their fundamental thinking, culture and practice.

Studies have demonstrated the effectiveness of business coaching to build the capacity of MSEs in a range of business practices (Blackman, 2008; Gray, Ekinci, & Goregaokar, 2011; Kutzhanova, Lyons, & Lichtenstein, 2009; Powell & Houghton, 2008), among which can be utilised for understanding and integrating tourism and sustainable development practices. Business coaching is a learning method that can be used to build capacity within a business to address its internal and external challenges. MSEs operating in challenging geographical areas, with limited access to a range of support services, face a number of challenges, requiring the necessary tools to analyse and make decisions.

Business coaching is a long-term management technique which is more comprehensive than other learning practices such as mentoring and job assignment, in terms of assessment, challenge and support because its ongoing nature seeks progressive achievement of goals. Business coaching helps focus the individual on particular goals through the use of one-on-one sessions which help with learning and behavioural change (Day, 2001; Hall et al., 1999; Peterson, 1996). It has also been suggested that business coaching is more effective than other forms of training and development. For example, some authors have argued that business coaching provides better transfer of training in the workplace than other techniques because it is grounded in the coachee's workplace (Bacon & Spear, 2003; Eggers & Clark, 2000; Joo, 2005; Olivero, Denise Bane, & Kopelman, 1997). There are several different approaches to coaching and any one coaching program can include a range of techniques and activities. A combination of techniques that provides maximum effectiveness to the coaching process is through the use of workshops and one-on-one coaching sessions (Damon, 2007).

In the following sections, the practice of business coaching as a capacity building tool and learning method for MSEs addressing the issues of sustainability in tourism will be presented.

10.5 Adapting Coaching to the MSE Context

While business coaching can be used to address barriers and build capacity in and around tourism and sustainability, there is a need to adapt the coaching to the specific circumstances of the MSEs for it to work effectively. Added to this are the additional challenges tourism MSEs encounter which also need to be taken into account when designing coaching programmes. Generally, challenges can include such aspects as, an understanding of environmental and socio-cultural impacts, the need for strong leadership, having access to local networks, understanding the tourism industry and how to get involved, and a lack of time and resources to invest in developing the business further.

Challenges facing tourism MSEs can be divided into internal and external. Internal challenges are associated with the internal operations of a business, while external challenges are extrinsic influences on the business, both of which affect the competitiveness, growth and profitability of MSEs. Blackman et al. (2004) identified a number of barriers for MSEs to successfully develop tourism in peripheral areas, which could be divided into internal and external from an MSE perspective. Internal challenges facing MSEs include difficulties with finance, a lack of education, resources or experience with the tourism sector, and lack of resources for training, marketing, implementing sustainable development practices etc. External challenges or barriers include a lack of control over negative impacts, a lack of economies of scale, community opposition, high fixed costs, a lack of infrastructure, limited market opportunities, and impacts of seasonality (Carson & Carson, 2011; Jackson & Murphy, 2006; Lordkipanidze et al., 2005; Macbeth et al., 2004).

Both internal and external challenges can be effectively addressed through the coaching process utilising different techniques to achieve the required outcomes. As previously stated, the use of coaching workshops where the MSE operators are able to learn skills and are provided the opportunity to network and use one-on-one follow-up coaching sessions is an effective way for businesses to deal with some of the challenges faced. Tourism MSEs should be nurtured through a series of stages in their business aspirations engaging in professional advice, and participating in training on key business aspects and leadership (Buhalis & Peters, 2006). Skills developed from business coaching helps to create an awareness of the types of barriers that exist, allowing MSEs to develop contingency plans for how to deal with the barriers. In addressing barriers at the regional level, techniques such as workshops are used widely for MSEs (Day, 2001), to allow for participants to interact and develop social capital through the use of networking. Workshops provide a conducive capacity building environment for MSE business owners as they are able to network with others in their region, enabling them to see past their colleagues as competitors and understand how to work together to best develop their destination to make a positive contribution to community wellbeing.

More specifically, internal challenges can be met by developing an MSE's individual business goals with a one-on-one coaching method, that provides a continual feedback process of goal redevelopment and achievement translating into action plans. This type of approach helps MSEs deal with some of the challenges faced from an industry and destination perspective that have largely been absent in their traditional industries. One reason for this is predominately contained within the nature of the traditional rural economy. Carson and Carson (2011) found that for many rural businesses, delivering single product raw materials to wholesalers at fixed or predetermined prices meant that they had no need to diversify their product or consider competition in the same way that businesses in tourism or other industries would have to. External challenges can be perceived to be more outside the ability of an individual MSE owner to address. In these circumstances the process of business coaching can provide regional MSE owners with the ability to form networks and provide leadership to overcome challenges and barriers.

Similarly, one of the specific themes for community leadership in the tourism sector identified by Moscardo (2005) is to have extensive local networks. In order for tourism MSEs to develop leadership abilities, coaching helps to facilitate leadership skills and provide the opportunities for important networks to be formed, allowing MSEs to move or integrate more easily into the tourism system. There has been evidence where networks are taking the role of a learning mechanism, towards understanding and integrating sustainable development into business practice. However rather than the actual network as a mechanism contributing to the achievement of sustainable tourism, it is the process of cooperation between network members that has been found to be more important (Halme, 2001). More importantly, it is said that while sustainable development goals are inextricably linked with the personal, business or destination benefits provided by a network, such benefits also have a significant motivational role in the process in which sustainable development is applied at a practical level (Halme & Fadeeva, 2000).

10.5.1 Adapting MSEs to the Tourism Sector

Participation in tourism by many MSEs is often limited due to a lack of understanding of the industry and how to enter the industry. For those MSEs wishing to develop or adapt their business to tourism, it is important to understand what skills are needed to set-up and operate tourism businesses, in addition to the importance of understanding those unique issues that are specific to the tourism industry. In understanding those issues Breen, Bergin-Seers, Jago, and Carlsen (2005) suggest that tourism industry risks are different to other industries due to issues of seasonality and the impact external shocks can have on the industry (e.g. airline strikes, terrorism). They also stated the uncertainty of the market to be another issue, as is its sensitivity to economic circumstances. Macionis (1997) found those in the wine industry adapting their business to tourism lacked experience in tourism knowledge and entrepreneurial skill which often prevented them from successfully entering the industry. Similarly Carson and Carson (2011) found that a lack of skills in tourism and a limited understanding of the interconnected complexities of a tourism destination system, was a major barrier for local business owners to develop competitive tourism products. They identified a lack of skills in the areas of tourism marketing, business administration, and customer service. With tourism tightly integrated into the delivery of services, human resource management and service delivery is also considered an important aspect in better understanding the tourism industry, particularly when compared with other industries (Breen et al., 2005). In rural settings where typically homogenous industries (agriculture, mining, steel) have dominated, the need to create market knowledge to develop competitive products can be largely absent, as is the need to undertake activities not previously considered in the business operations realm, such as marketing (Carson & Carson, 2011). In understanding and overcoming these issues unique to the industry, skills (tourism, entrepreneurial) and education/training were seen as critical for tourism MSEs in building capacity for competitive advantage (Buhalis & Peters, 2006).

When adapting generic education or coaching programmes for the tourism sector, Echtner (1995) found that the basic core components of the education process need to be retained, with industry specific cases and examples to be added, along with additional modules introduced to address unique industry issues. Gupta (1989) noted that in this industry, a specialised process is best accomplished by the use of tourism experts, such as government officials, academics and private sector business professionals. Breen et al. (2005) also considered that for education/training programmes in tourism, fundamentals for all types of businesses and sectors need to be the same; that is financial management, planning, networks or relationships, and knowing the market.

From a regional perspective, whole areas adapting to the tourism industry require a separate approach. For small tourism businesses to survive they must form or enter into cooperative bonds with others within the industry. Their competitive advantage is determined not by how efficiently they produce and market tourism products and services, but by their ability to "exploit available resources in the network" (Kelliher, Foley, & Frampton, 2009: 84). Local tourism leadership, local stakeholder coordination and participation (networking), and a lack of market information and tourism knowledge are some of the most common issues in developing a sustainable regional tourism sector (Kelliher et al., 2009; Moscardo, 2008). The key to tourism MSEs engaging successfully in the wider tourism region, requires the input of a range of different businesses to simultaneously cooperate as a group, whilst continuing to maintain individual competitive advantages (Jackson & Murphy, 2006).

Developing tourism in regional areas requires effective leaders who in turn need effective networks to achieve their goals. Leadership has been identified as a key factor in developing tourism in regional areas and one of the key features associated with effective tourism leaders was the access that they had to business networks (Long & Nuckolls, 1994; Teare, 1998; Wilson, Fesenmaier, Fesenmaier, & Van Es, 2001). The coaching process can be used to develop local leaders, and research into the role that leadership plays in successfully developing tourism in regional areas has been conducted in a wide range of countries including Australia (Kenyon & Black, 2001), Norway (Holmefjord, 2000), Portugal (Edwards & Fernandes, 1999), the Slovak Republic (Clarke, Denman, Hickman, & Slovak, 2001), the US (Lewis, 2001), Croatia (Petric, 2003) and Romania (Muica & Turnock, 2000).

10.6 Integration of Coaching and EfS for Tourism MSEs

In the context of this chapter, business coaching has been prescribed as a practical learning method to encourage sustainability of tourism MSEs. In the past, education or training for sustainability was delivered through a range of programmes that traditionally focused around environmental education. However, the concept of sustainable development has progressed to encompass a more broader definition of social, cultural and economic factors (Law, 2009).

It has been reviewed that a more targeted approach to delivering information to MSE owners or managers, supported by flexible, personalised programs for capacity building and personal support in combination with peer networks, could provide a more effective and sustainable approach to encourage MSEs to incorporate sustainability into their business practices. Tourism MSEs can improve the sustainability of their businesses when they are willing and able to cooperate with other tourism industry partners. This is where networking and leadership qualities of MSE owners is critical to the understanding of their contribution of sustainable development (Buhalis & Peters, 2006).

Moscardo in Chap. 1 of this book states that for education for sustainability to be effective it needs to take its cues from outside the formal education system, focusing on education as a "transformative experience for learners" where the objectives are more about capacity building and self-development, and where the teaching approaches are learner centred focusing on practice and action, and where systems and/or institutions curriculum and structure is organised by problems and skills where informal learning is practiced (based on Cotton & Winter, 2010; O'Brien et al., 2013; Rowe & Johnston, 2013; Ryan & Cotton, 2013; Sterling, 2004). In this respect, coaching is well positioned to take the lead and effectively assist in this transformation process of EfS.

If coaching, then, is to be effectively used to help tourism MSE operators, it is important to understand what an effective coaching process needs and how this integrates within the context of EfS. Initial activities in business coaching, such as the workshop, provide a key platform as the start of the coaching process to encourage participants to focus on their values and goals and to encourage a positive attitude towards change. This initial stage is valuable in helping participants to formulate detailed goals and translate these into action plans. The second stage of coaching, one-on-one sessions, allow for maintenance of change and further development of the MSE owner, where the coach's general skills and support have become more important than their technical expertise or sector experience. On another level, the use of the workshop provides participants with an opportunity to develop networks. Coaching adds value by providing the opportunity for MSE owners to interact with other operators through the exchange of ideas and a sense of place and pride in one's destination, and, therefore, contributing to the wellbeing of the MSE operator and the wider community. It is suggested that there are different phases or cycles that the MSEs will need to go through during the coaching process, cycling between group and individual work. The first cycle is where MSE owners learn how to adopt and adapt to tourism and sustainable development, which then cycles into the second cycle where MSEs come together to adopt sustainable development at the regional level through leadership and networks (see Fig. 10.1).

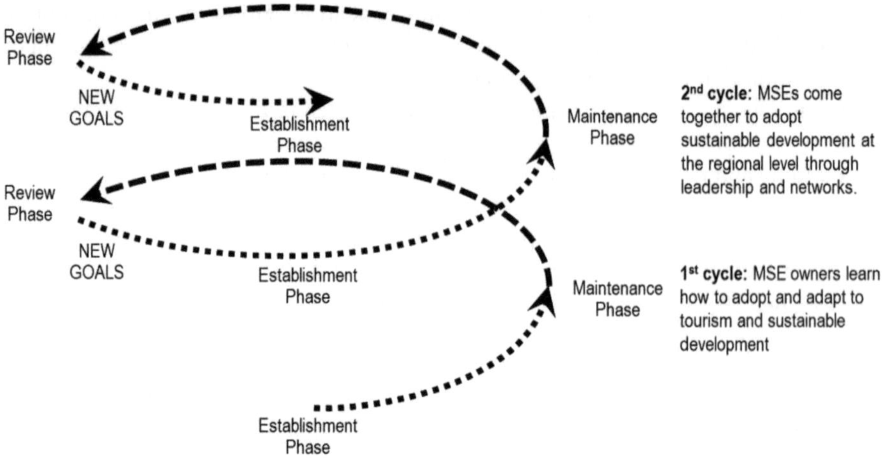

Fig. 10.1 Coaching sustainability development cycle

Phase 1: Workshop

The workshop provides a key platform as the start of the coaching process to encourage participants to focus on their values and goals and to encourage a positive attitude towards change. The skills learnt in the workshop allow the participants to develop capacity within their businesses, those that complete the one on one sessions find that they have been able to better implement the learned knowledge into their daily work routines. In this first phase tourism specific information is valuable in helping participants to formulate detailed goals and translate these into action plans. It also provides participants an opportunity to develop networks with others from the local region and promotes community connectedness.

Phase 2: One-on-One Sessions

The technique of follow up sessions with participants provides an opportunity to continuously revisit goals set in the workshop and adjust to the markets that influence them. They are also more confident in developing support networks. They become the epitome of the spiral of coaching—working towards one goal and as this is achieved they are able to start the coaching process again on a new goal but at a higher level; they are using the skills and knowledge gained to achieve their goals.

The one-on-one sessions then allow for maintenance of change and further development. In the maintenance phase the coach's general skills and support become more important than their technical expertise or sector experience. When participants are able to break large goals down into smaller more achievable steps (a learned coaching technique), it allows for the transfer of skills into the workplace. They also have a desire to engage in a wide range of development activities including more workshops. This suggests that an effective sustainable development approach could be a cycle between individual and group work (Damon, 2007) such as that set out in Fig. 10.1.

References

Bacon, T., & Spear, K. (2003). *Adaptive coaching: The art and practice of a client-centered approach to performance improvement*. Palo Alto, CA: Davies-Black.

Barton, C. (2010). *Understanding corporate social responsibility engagement in small and medium tourism businesses*. Thesis submitted in partial fulfilment of the requirements for the degree of Bachelor of Management (Honours) in Tourism, The University of Technology, Sydney.

Bauer, J., Sofield, T., Webb, J., Battig, M., & De Lacy, T. (n.d.). Conservation, poverty alleviation and community development through tourism in developing countries: The International Program of the CRC for Sustainable Tourism (1999-2002): Cooperative Research Centre for Sustainable Tourism.

Biggs, D., Hall, C. M., & Stoeckl, N. (2012). The resilience of formal and informal tourism enterprises to disasters—Reef tourism in Phuket. *Journal of Sustainable Tourism, 20*(5), 645–665.

Blackman, A. (2008). Perspectives on leadership coaching for regional tourism managers and entrepreneurs. In G. Moscardo (Ed.), *Building community capacity for tourism development* (pp. 142–154). Wallingford: CABI.

Blackman, A., Foster, F., Hyvonen, T., Jewell, B., Kuilboer, A., & Moscardo, G. (2004). Factors contributing to successful tourism development in peripheral regions. *The Journal of Tourism Studies, 15*(1), 59–70.

Bramwell, B. (1994). Rural tourism and sustainable rural tourism. *Journal of Sustainable Tourism, 2*(102), 1–6.

Breen, J., Bergin-Seers, S., Jago, L., & Carlsen, J. (2005). *Small and medium tourism enterprises: The identification of good practice*. Gold Coast, QLD: CRC for Sustainable Tourism.

Buhalis, D., & Peters, M. (2006). SMEs in tourism. In D. Buhalis & C. Costa (Eds.), *Tourism management dynamics* (pp. 116–129). Oxford, UK: Elsevier.

Carlsen, J., Getz, D., & Ali-Knight, J. (2001). The environmental attitudes and practices of family businesses in the rural tourism and hospitality sectors. *Journal of Sustainable Tourism, 9*(4), 281–297.

Carson, D. A., & Carson, D. B. (2011). Why tourism may not be everybody's business: The challenge of tradition in resource peripheries. *The Rangeland Journal, 33*(4), 373–383.

Clarke, J., Denman, R., Hickman, G., & Slovak, J. (2001). Rural tourism in Roznava Okres: A Slovak case study. *Tourism Management, 22*, 193–202.

Commonwealth of Australia. (2011). *Key statistics*. Canberra: Australian Small Business. Department of Innovation, Industry, Science and Research.

Cotton, D., & Winter, J. (2010). 'It's not just bits of paper and light bulbs': A review of sustainability pedagogies and their potential for use in higher education. In P. Jones, D. Selby, & S. Sterling (Eds.), *Sustainability education: Perspectives and practice across higher education* (pp. 39–54). Abingdon: Earthscan.

Damon, N. (2007). Follow the successful leader. *Training & Coaching Today*, 10–11.

Day, D. (2001). Leadership development: A review in context. *Leadership Quarterly, 11*(4), 581–613.

Denman, R. (1994). Green tourism and farming. In J. Fladmark (Ed.), Cultural tourism (pp. 215–222). Papers presented at the Robert Gordon University Heritage Convention. In Carlsen, J., Getz, D., & Ali-Knight, J. (2001). The environmental attitudes and practices of family businesses in the rural tourism and hospitality sectors. *Journal of Sustainable Tourism, 9*(4), 281–297.

Echtner, C. (1995). Entrepreneurial training in developing countries. *Annals of Tourism Research, 22*(1), 119–134.

Edwards, J., & Fernandes, C. (1999). Emigrants and espigueiros—Tourism activities in a peripheral area of Portugal. *International Journal of Tourism Studies, 1*(5), 329–340.

Eggers, J., & Clark, D. (2000). Executive coaching that wins. *Ivey Business Journal, 65*(1), 66–71.

Garrod, B., Warnell, R., & Youell, R. (2006). Re-conceptualising rural resources as countryside capital: The case of rural tourism. *Journal of Rural Studies, 22*, 117–128.

Gebremichael, B. A. (2014). The impact of subsidy on the growth of small and medium enterprises (SMEs). *Journal of Economics and Sustainable Development, 5*(3), 178–188.

Gibb, A. (2009). Meeting the development needs of owner managed small enterprise: A discussion of the centrality of action learning. *Action Learning: Research and Practice, 6*(3), 209–227.

Goodall, B., & Stabler, M. J. (1992). Environmental auditing in the quest for sustainable tourism: The destination perspective. In *Tourism in Europe. Conference proceedings* (pp. G1–G13). Newcastle Polytechnic.

Gray, D., Ekinci, Y., & Goregaokar, H. (2011). Coaching SME managers: Business development or personal therapy? A mixed methods study. *The International Journal of Human Resource Management, 22*(4), 863–882.

Gunn, C. A. (1988). *Tourism planning* (2nd ed.). New York: Taylor & Francis.

Gupta, S. K. (1989). Entrepreneurship development: The Indian case. *Journal of Small Business Management, 27*, 67–69. In Echtner, C. (1995). Entrepreneurial training in developing countries. *Annals of Tourism Research, 22*(1), 119–134.

Hall, D., Otazo, K., & Hollenback, G. (1999). Behind closed doors. *Organizational Dynamics, 27* (3), 39–53.

Halme, M. (2001). Learning for sustainable development in tourism networks. *Business Strategy and the Environment, 10*(2), 100–114.

Halme, M., & Fadeeva, Z. (2000). Small and medium-sized tourism enterprises in sustainable development networks. *Greener Management International, 30*, 97–113.

Holmefjord, K. (2000). *Synergies in linking products, industries and place? Is co-operation between tourism and food industries a local coping strategy in Lofoten and Hardanger.* Paper presented at the whether, how and why regional policies are working in concert with coping strategies locally workshop, Finland.

Jackson, J., & Murphy, P. (2006). Clusters in regional tourism. An Australian case. *Annals of Tourism Research, 33*(4), 1018–1035.

Joo, B. (2005). Executive coaching: A conceptual framework from an integrative review of practice and research. *Human Resource Development Review, 4*(4), 462–488.

Kelliher, F., Foley, A., & Frampton, A. M. (2009). Facilitating small firm learning networks in the Irish tourism sector. *Tourism and Hospitality Research, 9*, 80–96.

Kenyon, P., & Black, A. (2001). *Small town renewal: Overview and case studies* (Publication 01/043). Canberra: Rural Industries Research and Development Corporation.

Khan, M. R. (2013). Mapping entrepreneurship ecosystem of Saudi Arabia. *World Journal of Entrepreneurship, Management and Sustainable Development, 9*(1), 28–54.

Kutzhanova, N., Lyons, T. S., & Lichtenstein, G. A. (2009). Skill-based development of entrepreneurs and the role of personal and peer group coaching in enterprise development. *Economic Development Quarterly, 23*, 193–210.

Lane, B. (1994). Sustainable rural tourism strategies: A tool for development and conservation. *Journal of Sustainable Tourism, 2*(1&2), 102–111.

Law, B. (2009). Education for sustainable development in New Zealand. In B. Chalkley, M. Haigh, & D. Higgitt (Eds.), *Education for sustainable development: Papers in honour of the United Nations Decade of Education for Sustainable Development* (pp. 2005–2014). London: Routledge.

Lewis, J. (2001). Self-developed rural tourism: A method of sustainable tourism development. In S. McCool & R. Moisey (Eds.), *Tourism, recreation and sustainability: Linking culture and the environment* (pp. 177–194). Wallingford: CABI.

Long, P., & Nuckolls, J. (1994). Organising resources for rural tourism development: The importance of leadership, planning and technical assistance. *Tourism Recreation Research, 19*(2), 19–34.

Lordkipanidze, M., Brezet, H., & Backman, M. (2005). The entrepreneurship factor in sustainable tourism development. *Journal of Cleaner Production, 13*(8), 787–798.

Macbeth, J., Carson, D., & Northcote, J. (2004). Social capital, tourism and regional development: SPCC as a basis for innovation and sustainability. *Current Issues in Tourism, 7*(6), 502–522.

Macionis, N. (1997). Wine tourism in Australia: Emergence, development, and critical.

Moscardo, G. (2005). Peripheral tourism development: Challenges, issues and success factors. *Tourism Recreation Research, 30*(1), 27–43.

Moscardo, G. (2008). Sustainable tourism innovation: Challenging basic assumptions. *Tourism and Hospitality Research, 8*(1), 4–13.

Moscardo, G. (2011). The role of knowledge in good governance for tourism. In E. Laws, H. Richins, J. Agrusa, & N. Scott (Eds.), *Tourist destination governance* (pp. 67–80). Wallingford: CABI.

Muica, N., & Turnock, D. (2000). Maramures: Expanding human resources on the Romanian periphery. *GeoJournal, 50*, 181–198.

Murta, J., Gero, A., Kuruppu, N., & Mukheibir, P. (2012). Enhancing adaptive capacity of small to medium enterprises—Background Report (draft) (pp. 1–43). Prepared for National Climate Change Adaptation Research Facility, Institute for Sustainable Futures, University of Technology Sydney, Sydney, Australia.

Nuwagaba, A., & Nzewi, H. (2013). Major environmental constraints on growth of micro and small enterprises in Uganda: A survey of selected micro and small enterprises in Mbarara Municipality. *International Journal of Cooperative Studies, 2*(1), 26–33.

O'Brien, K., Reams, J., Caspari, A., Dugmore, A., Faghihimani, M., Fazey, I., et al. (2013). You say you want a revolution? Transforming education and capacity building in response to global change. *Environmental Science & Policy, 28*, 48–59.

Olivero, G., Denise Bane, K., & Kopelman, R. (1997). Executive coaching as a transfer of training tool: Effects on productivity in a public agency. *Public Personnel Management, 26*(4), 461–469.

Ongoro, Z. O., Kiragu, E. M., & Simwa, S. O. (2013). Effect of managerial skills on growth of hotel based micro and small enterprises in Keroka, Kenya. *Interdisciplinary Journal of Contemporary Research in Business*. Institute of Interdisciplinary Business Research, UK, 4(12), 370–374.

Peterson, D. (1996). Executive coaching at work: The art of one-on-one change. *Consulting Psychology Journal: Practice and Research, 48*(2), 78–86.

Petric, L. (2003). *Constraints and possibilities of the rural tourism development with special stress on the case of Croatia*. Paper presented at the European Regional Science Association 2003 Congress, Finland.

Powell, J. A., & Houghton, J. (2008). Action learning as a core process for SME business support. *Action Learning: Research and Practice, 5*(2), 173–184.

Rainforest Alliance. (2005). *Guide for sustainable tourism best practices* (2nd ed.). New York: Sustainable Tourism Division, Rainforest Alliance.

Roberts, S., & Tribe, J. (2008). Sustainability indicators for small tourism enterprises—An exploratory perspective. *Journal of Sustainable Tourism, 16*(5), 575–594.

Rowe, D., & Johnston, L. F. (2013). Learning outcomes: An international comparison of countries and declarations. In L. F. Johnston (Ed.), *Higher education for sustainability* (pp. 45–59). New York: Routledge.

Ryan, A., & Cotton, D. (2013). Times of change: Shifting pedagogy and curricula for future sustainability. In S. Sterling, L. Mazey, & H. Luna (Eds.), *The sustainable university* (pp. 151–167). Abingdon: Routledge.

Sharma, P., & Gounder, N. (2011). Obstacles to financing micro and small enterprises: Empirical evidence from a small island developing state. Griffith Business School Discussion Papers, (2011-10).

Sharpley, R. (2001). Sustainable rural tourism development: Ideal or idyll? In L. Roberts & D. Hall (Eds.), *Rural tourism and recreation: Principles to practice* (pp. 57–58). Wallingford: CAB.

Sterling, S. (2004). An analysis of the development of sustainability education internationally: Evolution, interpretation and transformative potential. In J. Blewitt & C. Cullingford (Eds.), *The sustainability curriculum* (pp. 43–62). London: Earthscan.

Teare, R. (1998). Enabling organisational learning. In R. Teare, J. Bowen, & N. Hing (Eds.), *New directions in hospitality and tourism: A worldwide review* (pp. 81–90). London: Cassell.

Tilbury, D. (2004). Rising to the challenge: Education for sustainability in Australia. *Australian Journal of Environmental Education, 20*(2), 103.

UNESCO. (n.d.). Education for sustainable development. Retrieved October 11, 2013, from http://www.unesco.org/new/en/education/themes/leading-the-international-agenda/education-for-sustainable-development/

Wilson, S., Fesenmaier, D., Fesenmaier, J., & Van Es, J. (2001). Factors for success in rural tourism development. *Journal of Travel Research, 40*(November), 132–138.

Chapter 11
Sustainability Education for Tourists

Gianna Moscardo

Abstract Tourists are a key stakeholder group that must be considered in any discussion of education for sustainability (EfS) in tourism. There are two main dimensions to this discussion—educating tourists to support the management of their impacts on the destination and using tourist experiences as an opportunity for EfS more broadly. This chapter identifies points of communication between tourism actors and tourists which could be used for EfS in tourism. It then reviews two main types of communication—heritage interpretation and destination marketing and outlines how these two processes will need to change to support EfS in tourism. Tourism could be an opportunity to educate people about global sustainability issues and encourage them to change their behaviours while at the destination, when they travel in the future and when they return to their everyday lives at home. To realise this opportunity will, however, require fundamental change in the processes of interpretation and destination marketing.

Keywords Interpretation • Marketing • Tourists • Visitor learning • Experiences

11.1 Introduction

Tourists are key stakeholders in any discussion of education for sustainability (EfS) in tourism. Krippendorf (1987, p. 43) argued that "*the damage tourism causes to the people, economy and environment of the host area, especially in the long-term, remain hidden from the tourist. He has been left out of all discussion on the subject, even though he is one of the main protagonists. . .They are therefore carefree and ignorant rather than devious. To lay all blame at their door would be as wrong as denying their responsibility. But they should certainly be [made] aware of the situation.*" In this argument Krippendorf acknowledges that while tourists are central to managing the impacts of tourism, they have not been given a voice in discussions of either tourism impacts or tourism sustainability. Moscardo (1998, p. 4) notes that "*the information they [tourists] encounter while at leisure may offer the only opportunity to learn about their bonds to the environment, or to their*

G. Moscardo (✉)
James Cook University, Townsville, QLD, Australia
e-mail: Gianna.moscardo@jcu.edu.au

G. Moscardo, P. Benckendorff (eds.), *Education for Sustainability in Tourism*, CSR, Sustainability, Ethics & Governance, DOI 10.1007/978-3-662-47470-9_11

history and culture". Taken together these two quotes identify two main dimensions of education for sustainability (EfS) in tourism—the need to educate tourists to minimise their negative impacts while travelling and the opportunity that tourism offers to provide educational experiences that could support sustainability in tourism and beyond.

This chapter will begin by identifying the main points of communication between different actors in the tourism system and tourists and use these points to establish where education relevant to sustainability in tourism exists now within the tourism system and to suggest further opportunities for EfS within tourism. It will then critically review the concepts and processes of the two main current approaches to communicating with tourists—interpretation and marketing. It will note some of the limitations and critiques of these two areas before suggesting some new directions based on concepts from social marketing that could support more extensive and more effective EfS in tourism. It will conclude by highlighting some challenges in moving forward in these new directions.

11.2 Communicating with Tourists: An Overview of Options

Tourism has often been linked to learning and education. According to Falk, Ballantyne, Packer, and Benckendorff (2012) learning is an important travel motivation and tourism offers a potentially important opportunity for education through intensive first hand experiences. Despite this there has been little systematic attention paid to tourist learning by tourism researchers or practitioners with the exception of work in the field of interpretation (Falk et al., 2012). This chapter seeks to understand tourist learning in order to support EfS in tourism. To explore this potential for tourism to support learning it is useful to understand all the points of communication between the different actors in the tourism system. Figure 11.1 summarises these into nine main connections.

Destination marketing organisations offer a natural starting point as these are the groups that seek to communicate information about their destinations in order to raise awareness of the tourism possibilities in their regions and to encourage tourists to select their location and visit (Pike & Page, 2014). These marketing communications not only convey information about the nature of tourism products that are available in the focus destination, they also often include information about the history, culture and environmental features of the destination (Pike, 2008). Tourism marketing is also conducted by tourism businesses and organisations responsible for managing tourist attractions such as historic trusts and national park agencies, the second point of communication contact in Fig. 11.1. Tourism businesses and attraction managers need to communicate with tourists to provide information about the specific details of the products or services they offer including planning information, such as times and required equipment, and safety messages. Tourist

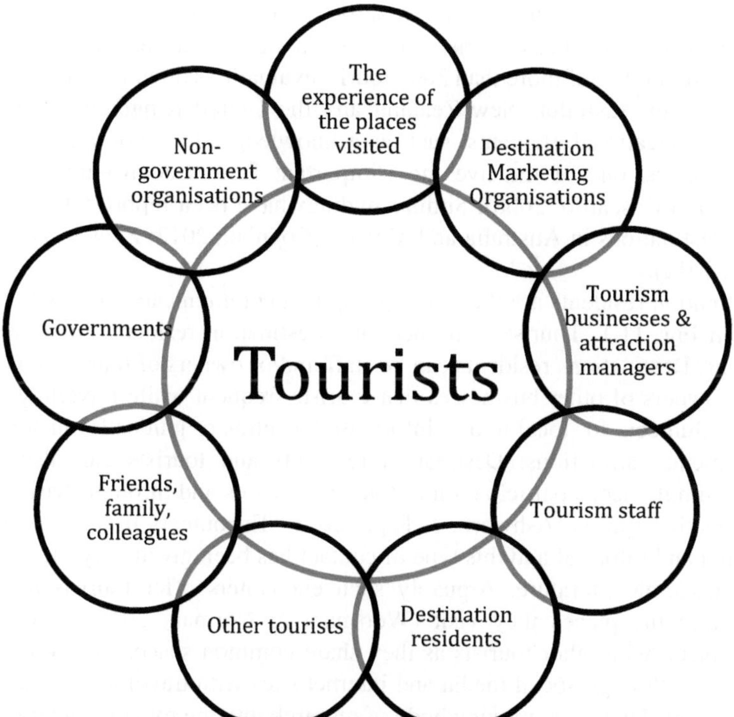

Fig. 11.1 Points of communication contact with tourists

businesses seeking to implement sustainability strategies must also communicate information about these to tourists and encourage them to support the strategies (Needham & Little, 2013; Teng, Horng, Hu, Chien, & Shen, 2012). This can, however, be a challenging task. Lee and Moscardo's (2005) research found that the rate of participation of guests in the various sustainability programs offered at an ecotourism resort in Australia was low even though many reported that they chose the resort because of its green accreditation. More recent research into the environmentally responsible behaviour of people at home and on holidays provides similar results. Acting in an environmentally responsible manner at home does not translate into acting in an environmentally responsible manner while on holidays (Barr & Prillwitz, 2012; Miao & Wei, 2013). This reflects a fundamental feature of taking a holiday or being a tourist; which is to escape everyday routines and responsibilities (Cohen, Higham, & Reis, 2013).

Staff in tourism businesses and organisations play a major role in conveying information to their guests. This can happen informally during routine service interactions such as ticket sales, or more formally through guided tours and presentations. These educational activities focussed on explaining the historical, cultural and/or environmental significance of the places being accessed with the tourism business or organisation are a core feature of many tourist experiences and are often referred to as interpretation. This interpretation is not only an expected feature for many tourists, it is also a significant contributor to tourism satisfaction. Studies of more than 3000 tourist evaluations of wildlife based tourism experiences in Australia, New Zealand and the United Kingdom, for example, identified educational elements, such as a knowledgeable guide and educational information, as one of the five most important factors contributing to overall satisfaction (Moscardo, 2008). Similar findings have been reported for tourists to coastal destinations in Australia and Mexico (Coghlan, 2012; Ziegler, Dearden, & Rollins, 2012).

Destination residents are the fourth group that communicate with tourists in the model in Fig. 11.1. Tourists can encounter destination residents in a number of situations. Destinations residents can be staff and/or owners of tourism businesses, staff or owners of other businesses that tourists frequent while travelling, or they can be volunteers for local tourist information centres or guides for historical and environmental attractions. Destination residents and tourists can also interact through simultaneous participation in tourist activities and through sharing leisure and recreation spaces (Ashworth & Page, 2011). Encounters in these settings are incidental and informal and this type of contact has been given very little attention in the academic literature. Arguably such encounters offer tourists a different insight into the places they visit (Woosnam & Norman, 2010). Tourists also communicate with other tourists as they share common spaces and activities and increasingly through social media and internet sites with travel reviews and travel question fora. There is a growing body of research into the role that other travellers play in providing information about destinations and tourist businesses that can assist tourist decisions (Xiang & Gretzel, 2010). This inter-tourist communication also offers an opportunity for tourists to learn about other cultures, but has been given very little research attention (Nicholls, 2011).

The sixth group in Fig. 11.1 are the family, friends and colleagues of tourists. This group is another important source of information about places and activities that tourists can and do use to make decisions (Weaver & Lawton, 2011). To date research attention has only be paid to the information communicated from family, friends and colleagues to tourists, usually before they leave. What has not yet been studied in detail is the information about the destinations that tourists communicate back to their family, friends and colleagues as they travel and when they return. While there is research exploring how tourists report on their travels, especially through social media and the internet, it has focussed mostly on the structure of the information and how it serves the motives of the tourists (Yoo & Gretzel, 2011). Little, if any, research appears to have explored what the audiences of these travel stories learn from this communication. The potential role of tourists as educators about places and cultures when they return home has not yet been given serious consideration.

Various government agencies communicate with tourists. Here is there some overlap with destination marketing organisation as many governments run tourist promotion agencies. There is also some overlap with tourist businesses and attractions with government owned airlines and travel companies and government managed historical, cultural and environmental attractions. Beyond these roles governments also communicate both with their own citizens as tourists going elsewhere and with tourists from other regions or countries entering their jurisdictions. Information here is focussed on issues such as visas, customs and quarantine information, and health and safety management. Many countries offer travel advisories and travel safety information. The Australian government Smart traveller website (www.smartraveller.gov.au), for example, provides information for each country on overall safety, the level of potential terrorist threat, visa and entry requirements, levels of crime, transport safety, local laws, health issues, and the quality of medical facilities. Governments sometimes include information to tourists on how to behave in culturally appropriate ways while travelling. The Chinese government, for example, has issued a set of guidelines for civilized travel abroad to its citizens in a response to complaints from other countries about the inappropriate behaviours reported for some Chinese travellers (Johanson, 2013). These examples reflect awareness about the potential for negative cultural impacts arising from tourist behaviour. To date guidelines for being a responsible tourist beyond just the cultural dimensions have mostly been provided by non-government organisations such as Responsible Travel, Tourism Concern and Ethical Escape. These types of programs offer one way to educate tourists about the sustainability of their actions. This pathway to support sustainability in tourism faces, however, the major challenge of tourist awareness with research generally suggesting that few tourists are aware of, or use, this information (Marchoo, Butcher, & Watkins, 2014).

The final element of the model in Fig. 11.1 is the experience of the destination place itself. It is often assumed that tourists can learn about themselves and their place in the world through the kind of immediate and intensive experiences offered by tourism (Falk et al., 2012). There is some evidence to support this assumption (cf., Ballantyne, Packer, & Sutherland, 2011; Reijnders, 2011; Sharpley & Jepson, 2011; Walker & Moscardo, 2014), although the extent to which this type of experience contributes to changing tourists' understanding of key dimensions relevant to sustainability has not yet been examined in detail. In particular it is not clear what, if anything, tourists learn in the absence of guidance, preparation or interpretation. While some commentators have argued that it is important for tourists to have direct experience of places without such guidance or interpretation (O'Toole, 2007), the available research suggests that many tourists need support from suitably trained interpretive staff to help them to understand what they are experiencing and make sense of it (Walker & Moscardo, 2014).

This review of the points of communication between different actors and elements of the tourism systems reveals two main features. Firstly, it identifies a number of potential opportunities to improve the role of tourism as a tool for EfS both within the tourism context and beyond it. These include actions such as government and destination marketing organisation support for presenting responsible tourism guidelines, a focus on preparing tourists to be educators about the destinations they visit both while they travel and when they return home, and an exploration of providing more positive interactions between tourist and destination residents. Secondly, it highlights the critical role of interpretation and marketing in any attempts to provide EfS in tourism.

11.3 Traditional Approaches to Tourist Interpretation

Interpretation is the communication aimed at tourists that is most explicitly concerned with education and the sustainability of tourism. According to Tilden (1977, p. 33) interpretation can be defined as "*an educational activity which aims to reveal meanings and relationships through the use of original objects, by first hand experience, and by illustrative media*". Moscardo (2001) reviewed a range of definitions of interpretation and concluded that the most common concepts associated with interpretation included education, explanation, discovery, awareness, enjoyment and inspiration. Interpretation has mostly been used in tourism to enhance tourist experiences and to manage tourists' on-site behaviours in order to minimise negative impacts. Interpretation can be used to manage tourist behaviours directly through controlling or influencing where tourists can go on a site, filtering out those with less interest in all aspects of the site from more fragile areas, distributing visitors in ways that alleviate pressure on the place, raising awareness of possible negative impacts, and directly explaining the need for, and details of, guidelines and rules designed to minimise negative impacts (Bramwell & Lane, 2005; Kim, Airey, & Szivas, 2011; Moscardo, 1998, 2001). Interpretation can also work to manage tourism impacts indirectly through enhancing tourist understanding of the place and its significance and supporting an enjoyable or satisfactory experience and through these developing a sense of place attachment which encourages minimal impact behaviour on site and support for conservation of the place more broadly (Bramwell & Lane, 2005; Kim et al., 2011; Moscardo, 1998, 2001; Ramkissoon, Smith, & Weiler, 2013).

There is consistent evidence that interpretation can enhance visitor experiences and contribute to satisfaction (Biran, Poria, & Oren, 2011; Hughes, Packer, & Ballantyne, 2011; Moscardo, 2009a, 2009b; Walker & Moscardo, 2014). The evidence that this experience enhancement is linked to behaviour change or that interpretation is effective at increasing tourists' awareness and understanding of impacts, or at changing their behaviour is much less consistent (Littlefair & Buckley, 2008). Reviews conducted by Wiener, Needham, and Wilkinson (2009) and Halpenny (2010) concluded that the evidence for interpretation effectiveness is mixed at best. Even those who argue that interpretation can be effective in all these dimensions acknowledge that there is little consistent evidence of behaviour change beyond the setting or to support many of the assumed links between changes in attitudes and changes in behaviours or between behavioural intentions and actual behaviours (Hughes, 2013; Hughes et al., 2011; Powell & Ham, 2008). Munro, Morrison-Saunders, and Hughes (2008) go further providing a detailed critique of the methodological weaknesses common to many interpretation evaluations casting further doubt on its effectiveness.

In the context of EfS in tourism this brief review of interpretation suggests it may have only limited potential as a tool for educating tourists to support sustainability beyond the specific destination. Such a conclusion is, however, overly pessimistic as it fails to take into account the fact that interpretation has to date been mostly concerned with the specific place being interpreted and has not aimed to change understanding and behaviour beyond the specific setting. It is clear that not all interpretation is well-designed and in many situations the interpretation is isolated from other aspects of the tourist experience. There is an increasing awareness of the potential to use tourism as an opportunity to present messages about sustainability more generally with more attention being paid to the use of tourist interpretation to address broad sustainability messages. Blewitt (2013), for example, argues that places such as the Eden project in Cornwall and the Centre for Alternative Technology in Wales are potentially very effective at EfS. Research reported by Hughes, Packer, and Ballantyne (2011), Hughes (2013), and Powell and Ham (2008) on situations where there was a specific focus on sustainability beyond the tourist destination provides some evidence that it is possible to use tourism as a tool for EfS.

To achieve these goals interpretation has to be well designed, focussed on broader sustainability issues and integrated into the whole tourist experience. Features that have been shown to contribute to effective interpretation include:

- Providing variety across the physical and cognitive dimensions of the experience;
- Finding connections to the immediate personal experiences of the tourists;
- Examining the topic from multiple perspectives;
- Asking visitors to answer questions and giving them an active role in the interpretation;
- Using strong clear themes and structuring the information around stories;
- Highlighting the authenticity of the experience;
- Directing tourists to reflect on and consider their own values;
- Presenting consistent sustainability messages;
- Seeking their help as conservation partners;
- Providing follow-up contact and resources and information that can be accessed when they return home;
- Addressing barriers to change and providing clear instructions about the desired behaviour change; and
- Demonstrating wider social support for the desired behaviour changes (Ballantyne, Packer, & Hughes, 2009; Hughes, 2013; Kim et al., 2011; Moscardo, 2009a, 2009b).

Effective interpretation needs to encourage mindful tourists (Moscardo, 2009a, 2009b) who engage in reflective engagement about the subject being presented (Ballantyne, Packer, & Falk, 2011). It is important to note that many of the listed features are consistent with the characteristics of effective EfS more generally, as described in Chap. 1. In order to contribute to EfS interpretation also has to be supported by, and consistent with other, aspects of the tourist experience. It has been suggested that this requires a shift in tourism marketing approaches (Lai, Sorice, Nepal, & Cheng, 2009; Walker & Moscardo, 2014).

11.4 Tourism Marketing to Sustainability Marketing

Tourism marketing can be defined as a "continuous, sequential process through which management plans, researches, implements, controls and evaluates activities designed to meet customers' needs and wants and their own organizations' objectives" (Morrison, 2013, p. 9). There are two key features to this definition. Firstly, it is holistic and clearly describes tourism marketing as being about all aspects of tourism and its management and not just about advertising or selling tourist experiences. Secondly it includes the idea that organizations generally respond to consumer demand. Critiques of marketing argue that does more than just respond to consumer demand, rather traditional marketing seeks to encourage unnecessary and excessive consumption in order to support increased sales of goods and services (Jack et al., 2011). Other critiques of traditional marketing are that it is focussed on the specific product or service being sold and fails to consider the whole lifecycle of the product or service and thus ignores many of the sustainability issues associated with consumption (van Dam & Apeldoorn, 1996). Further, traditional marketing often fails to balance short-term individual satisfaction of wants against longer-term needs and the well-being of the collective (Belz & Peattie, 2010). Arguably tourism marketing can be seen as vulnerable to these same challenges.

In response to these challenges a number of different forms of, or approaches to, marketing have been proposed including environmental, societal, social, and sustainable marketing (see Belz & Peattie, 2010; Jamrozy, 2007; Truong & Hall, 2013 for reviews). Belz and Peattie (2010) offer the most comprehensive approach arguing that sustainability marketing has the following core principles:

- Supports a business that is economically viable and competitive;
- Is ethical and ecologically oriented;
- Based on equitable relationships with all key stakeholders;
- Balances consumers' rights and responsibilities with a focus on needs rather than wants;
- Takes a longer-term, total lifecycle approach and considers a wide range of impacts from the production, consumption and disposal of goods and services;
- Incorporates more than just economic objectives, but also ecological, social and sustainability objectives;
- Is integrated into relevant corporate social responsibility strategies;
- Incorporates education for consumers about sustainability;
- Includes social marketing campaigns; and
- Evaluates the effectiveness of the strategy by including assessments of both positive and negative impacts for the key stakeholders.

11.5 Implications for Encouraging Effective EfS with Tourists

Applying these principles to tourism marketing and interpretation in order to support sustainability in tourism and develop the role of tourism as a tool for EfS suggests five major changes to current practices are required. The first is to examine actual and potential tourist markets not just in terms of likelihood of visiting and revenue value for tourism businesses, but also to include assessments of the distance and type of transport used and its impacts, their likely behaviours in the destination and the impacts of these and their potential willingness to accept and participate in sustainability programs at the destination (Dinan & Sargeant, 2000). The second is a need to assess potential tourism products more thoroughly in terms of all possible impacts including social, cultural and ethical considerations as well as ecological impacts (Pomering, Noble, & Johnson, 2011) and to ensure that the product and marketing themes are consistent with destination community values and perceptions (Walker & Moscardo, 2014). The third is the need to measure tourism success in making an overall positive contribution to all aspects of the quality of life of the individual tourist and of the destination community (Jamrozy, 2007). The fourth change is to include tourist education about sustainability, both at the destination and beyond it, into both destination marketing and interpretation. The kind of consumer education described by Belz and Peattie (2010) is similar to that described in the previous section on interpretation. The key difference between the two is the need to educate tourists about the impacts of all aspects of their travel, and not just their behaviour at specific heritage sites. Finally there is a need to integrate destination marketing and interpretation to incorporate social marketing into the overall destination experience. Social marketing is "*a process that applies marketing principles and techniques to create, communicate, and deliver value in order to influence target audience behaviors that benefit society as well as the target audience*" (Cheng, Kotler, & Lee, 2011, p. 2). Social marketing is about using the marketing process to change behaviours. Given the need to change tourist behaviours is a core condition for improving the sustainability of tourism, it is surprising that it has not been given much attention to date in the tourism literature (Truong & Hall, 2013).

Figure 11.2 summarises the main areas of education or communication that need to be added to current approaches to interpretation and tourism marketing to maximize EfS for tourists. It suggests that education for tourists needs to be organised around three key themes—how to act sustainably within the destination, how to act sustainably as a tourist beyond the destination and how to act sustainably when they return home. The current attempts at tourist education are those highlighted as circles within the destination with interpreters focussing on minimal impact behaviours only at specific heritage sites and tourism marketers only focussing on information to encourage the purchase of tourism products within the destination. In order to support sustainability in tourism both heritage interpreters and destination marketers need to expand their approach to develop and

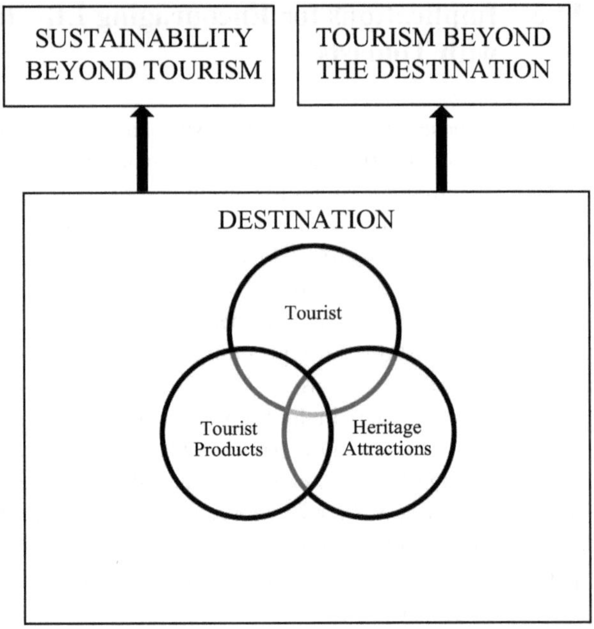

Fig. 11.2 Areas for expansion for interpretation and tourism marketing

present consistent information to tourists on how act in a sustainable fashion while in the destination as a whole. In order to address the second theme of tourism beyond the destination, education within the destination needs to include knowledge about tourism impacts more widely and support tourists to make choices about their future travel that are more sustainable. In order to address the third theme of sustainability beyond tourism, education within the destination needs to use destination experiences to highlight key sustainability issues and provide knowledge and support for adopting more sustainable lifestyles while at home. The latter means a change in the way interpreters think about heritage resources. For example, a guided tour on a coral cay is likely to focus on how the cay was created, the wildlife it supports and its importance in the ecosystem. Taking an EfS approach means adding to this approach by discussing climate change and its likely impact on the site and using that as a way to stimulate discussion and consideration of the behaviour changes needed to avoid these impacts. In the case of cultural heritage, a guided tour of a battlefield is likely to present details of the battle and an analysis of why it happened and what the consequences were. Taking an EfS approach suggests using the site visit as an opportunity for tourists to think about conflict over values and how they would respond to protect their own values and if there are better ways than war to deal with value conflicts.

11.6 EfS and Tourists: Challenges and Conclusions

While the adoption of social marketing and sustainability marketing principles and the development of an EfS focus in tourist experiences have potential to contribute to sustainability, these are not likely to happen in the short or even medium term. The type of change required to support EfS with tourists presents a number of challenges. Firstly, it requires a fundamental shift in objectives for both interpreters and marketers. The central objective for interpreters has been the preservation and presentation of heritage resources, both cultural and environmental. The central objective for marketers has been tourism growth. Currently there is no pressure from any stakeholder group for these objectives to be made secondary to sustainability. Destination marketers also face the issue of managing multiple actors in the tourism system. There are few effective models for consistent cooperation amongst all the businesses, organisations and individuals who are involved in the tourist experience. Thirdly, EfS with tourists will require a closer integration of marketing and interpretation and these are two groups with very different values and only limited understanding of each other. Destination marketers see heritage as a resource to exploit to attract tourists, while interpreters see heritage as an asset to be protected from exploitation.

These two groups do share one common assumption—that travel is a good thing. For marketers it represents a valuable exchange relationship, for interpreters it is an opportunity to learn and appreciate the world. In many places both groups are dependent on existing or growing numbers of tourists to support their activities. This shared dependence on tourism maintenance or growth leads to the final major challenge for EfS with tourists—the need to communicate a message about travelling and consuming less. Pomering et al. (2011) note that the ecological footprint of tourism continues to grow and with predicted increases in tourism from the so-called BRIC (Brazil, Russia, India, and China) (UNWTO, 2013) this is likely to become worse. Ultimately EfS in tourism needs to convince tourists to stay at home or closer to home and to find alternative ways to meet the needs they currently pursue with travel (Moscardo, 2009b).

References

Ashworth, G., & Page, S. J. (2011). Urban tourism research: Recent progress and current paradoxes. *Tourism Management, 32*(1), 1–15.

Ballantyne, R., Packer, J., & Falk, J. (2011). Visitors' learning for environmental sustainability: Testing short-and long-term impacts of wildlife tourism experiences using structural equation modelling. *Tourism Management, 32*(6), 1243–1252.

Ballantyne, R., Packer, J., & Hughes, K. (2009). Tourists' support for conservation messages and sustainable management practices in wildlife tourism experiences. *Tourism Management, 30* (5), 658–664.

Ballantyne, R., Packer, J., & Sutherland, L. A. (2011). Visitors' memories of wildlife tourism: Implications for the design of powerful interpretive experiences. *Tourism Management, 32*(4), 770–779.

Barr, S., & Prillwitz, J. (2012). Green travellers? Exploring the spatial context of sustainable mobility styles. *Applied Geography, 32*, 798–809.

Belz, F. M., & Peattie, K. (2010). *Sustainability marketing: A global perspective*. London: Wiley.

Biran, A., Poria, Y., & Oren, G. (2011). Sought experiences at (dark) heritage sites. *Annals of Tourism Research, 38*(3), 820–841.

Blewitt, J. (2013). *The ecology of learning: Sustainability, lifelong learning and everyday life*. Hoboken, NJ: Taylor & Francis.

Bramwell, B., & Lane, B. (2005). Interpretation and sustainable tourism: The potential and pitfalls. *Revista Interamericana de Ambiente y Turismo, 1*(1), 20–27.

Cheng, H., Kotler, P., & Lee, N. D. (2011). Social marketing for public health: An introduction. In H. Cheng, P. Kotler, & N. D. Lee (Eds.), *Social marketing for public health: Global trends and success stories* (pp. 1–30). Sudbury: Jones and Bartlett.

Coghlan, A. (2012). Facilitating reef tourism management through and innovative importance-performance analysis method. *Tourism Management, 33*(6), 767–775.

Cohen, S. A., Higham, J. E. S., & Reis, A. R. (2013). Sociological barriers to developing sustainable discretionary air travel behaviour. *Journal of Sustainable Tourism, 21*(7), 982–998.

Dinan, C., & Sargeant, A. (2000). Social marketing and sustainable tourism—Is there a match? *International Journal of Tourism Research, 2*(1), 1–14.

Falk, J. H., Ballantyne, R., Packer, J., & Benckendorff, P. (2012). Travel and learning: A neglected tourism research area. *Annals of Tourism Research, 39*(2), 908–927.

Halpenny, E. (2010). Pro-environmental behaviours and park visitors: The effect of place attachment. *Journal of Environmental Psychology, 30*, 409–421.

Hughes, K. (2013). Measuring the impact of viewing wildlife: Do positive intentions equate to long-term changes in conservation behaviour? *Journal of Sustainable Tourism, 21*(1), 42–59.

Hughes, K., Packer, J., & Ballantyne, R. (2011). Using post-visit action resources to support family conservation learning following a wildlife tourism experience. *Environmental Education Research, 17*(3), 307–328.

Jack, G., Higgins, M., Fitchett, J., Ellis, N., Lim, M., Tadajewski, M., et al. (2011). *Marketing: A critical textbook*. London: Sage.

Jamrozy, U. (2007). Marketing of tourism: A paradigm shift toward sustainability. *International Journal of Culture, Tourism and Hospitality Research, 1*(2), 117–130.

Johanson, N. (2013, October 2). China's 'guidelines on civilized travel abroad' offers these 25 pearls of wisdom. *IBT*. http://www.ibtimes.com/chinas-guidelines-civilized-travel-abroad-offers-these-25-pearls-wisdom-1414236. Accessed 5 May 2014.

Kim, A. K., Airey, D., & Szivas, E. (2011). The multiple assessment of interpretation effectiveness: Promoting visitors' environmental attitudes and behavior. *Journal of Travel Research, 50* (3), 321–334.

Krippendorf, J. (1987). *The holidaymakers: Understanding the impacts of leisure and travel*. London: William Heinemann.

Lai, P. H., Sorice, M. G., Nepal, S. K., & Cheng, C. K. (2009). Integrating social marketing into sustainable resource management at Padre Island national seashore: An attitude-based segmentation approach. *Environmental Management, 43*(6), 985–998.

Lee, W. H., & Moscardo, G. (2005). Understanding the impact of ecotourism resort experiences on tourists' environmental attitudes and behavioural intentions. *Journal of Sustainable Tourism, 13*(6), 546–565.

Littlefair, C., & Buckley, R. (2008). Interpretation reduces ecological impacts of visitors to world heritage site. *AMBIO: A Journal of the Human Environment, 37*(5), 338–341.

Marchoo, W., Butcher, K., & Watkins, M. (2014). Tour booking: Do travelers respond to tourism accreditation and codes of ethics initiatives? *Journal of Travel & Tourism Marketing, 31*(1), 16–36.

Miao, L., & Wei, W. (2013). Consumers' pro-environmental behaviour and the underlying motivations: A comparison between household and hotel settings. *International Journal of Hospitality Management., 32*, 102–112.
Morrison, A. M. (2013). *Marketing and managing tourism destinations*. Abingdon: Routledge.
Moscardo, G. (1998). Interpretation and sustainable tourism: Functions, examples and principles. *Journal of Tourism Studies, 9*(1), 2–13.
Moscardo, G. (2001). Cultural and heritage tourism: The great debates. In B. Faulkner, G. Moscardo, & E. Laws (Eds.), *Tourism into the twenty first century: Reflections on experience* (pp. 3–17). London: Cassells Academic.
Moscardo, G. (2008). Understanding visitor experiences in captive, controlled and noncaptive wildlife-based tourism settings. *Tourism Review International, 11*, 213–233.
Moscardo, G. (2009a). Understanding tourist experience through mindfulness theory. In M. Kozak & A. Decrop (Eds.), *Handbook of tourist behavior: Theory and practice* (pp. 99–115). New York: Routledge.
Moscardo, G. (2009b). Tourism and quality of life: Towards a more critical approach. *Tourism and Hospitality Research, 9*(2), 159–170.
Munro, J. K., Morrison-Saunders, A., & Hughes, M. (2008). Environmental interpretation evaluation in natural areas. *Journal of Ecotourism, 7*(1), 1–14.
Needham, M. D., & Little, C. M. (2013). Voluntary environmental programs at an alpine ski area: Visitor perceptions, attachment, value orientations, and specialization. *Tourism Management, 35*, 70–81.
Nicholls, R. (2011). Customer to customer interaction (CCI): A cross-cultural perspective. *International Journal of Contemporary Hospitality Management, 23*(2), 209–223.
O'Toole, F. (2007). Taming the Cliffs of Moher. *The Irish Times*. http://www.irishtimes.com/newspaper/archive/2007/0206/Pg016.html#Ar01602. Accessed 11 Sept 2013.
Pike, S. (2008). *Destination marketing*. Oxford: Butterworth-Heinemann.
Pike, S., & Page, S. J. (2014). Destination marketing organizations and destination marketing: A narrative analysis of the literature. *Tourism Management, 41*, 202–227.
Pomering, A., Noble, G., & Johnson, L. W. (2011). Conceptualising a contemporary marketing mix for sustainable tourism. *Journal of Sustainable Tourism, 19*(8), 953–969.
Powell, R. B., & Ham, S. H. (2008). Can ecotourism interpretation really lead to pro-conservation knowledge, attitudes and behaviour? Evidence from the Galapagos Island. *Journal of Sustainable Tourism, 16*(4), 467–489.
Ramkissoon, H., Smith, L. D., & Weiler, B. (2013). Testing the dimensionality of place attachment and its relationships with place satisfaction and pro-environmental behaviours: A structural equation modelling approach. *Tourism Management, 36*, 552–566.
Reijnders, S. (2011). Stalking the count: Dracula, fandom and tourism. *Annals of Tourism Research, 38*(1), 231–248.
Sharpley, R., & Jepson, D. (2011). Rural tourism: A spiritual experience? *Annals of Tourism Research, 38*(1), 52–71.
Teng, C.-C., Horng, J.-S., Hu, M., Chien, L.-H., & Shen, Y.-C. (2012). Developing energy conservation and carbon reduction indicators for the hotel industry in Taiwan. *International Journal of Hospitality Management, 31*(1), 199–208.
Tilden, F. (1977). *Interpreting our heritage* (3rd ed.). Chapel Hill: North Carolina Press.
Truong, V. D., & Hall, C. M. (2013). Social marketing and tourism what is the evidence? *Social Marketing Quarterly, 19*(2), 110–135.
UNWTO. (2013). International tourism on the rise boosted by strong performance in Europe. http://www2.unwto.org/press-release/2013-10-17/international-tourism-rise-boosted-strong-performance-europe-0. Accessed 12 May 2014.
van Dam, Y. K., & Apeldoorn, P. A. C. (1996). Sustainable marketing. *Journal of Macromarketing, 16*(2), 45–56.
Walker, K., & Moscardo, G. (2014). Encouraging sustainability beyond the tourist experience: Ecotourism, interpretation and values. *Journal of Sustainable Tourism, 22*(8), 1175–1196.

Weaver, D. B., & Lawton, L. J. (2011). Information sources for visitors' first awareness of a low profile attraction. *Journal of Travel and Tourism Marketing, 28*(1), 1–12.

Wiener, C. S., Needham, M. D., & Wilkinson, P. F. (2009). Hawaii's real life marine park: Interpretation and impacts of commercial marine tourism in the Hawaiian Islands. *Current Issues in Tourism, 12*(5–6), 489–504.

Woosnam, K. M., & Norman, W. C. (2010). Measuring residents' emotional solidarity with tourists: Scale development of Durkheim's theoretical constructs. *Journal of Travel Research, 49*(3), 365–380.

Xiang, Z., & Gretzel, U. (2010). Role of social media in online travel information search. *Tourism Management, 31*(2), 179–188.

Yoo, K.-H., & Gretzel, U. (2011). Influence of personality on travel-related consumer-generated media creation. *Computers in Human Behaviour, 27*(2), 609–621.

Ziegler, J., Dearden, P., & Rollins, R. (2012). But are tourists satisfied? Importance- performance analysis of the whale-shark tourism on Isla Holbox, Mexico. *Tourism Management, 33*(3), 692–702.

Part IV
Practices and Tools for Education for Sustainability in Tourism

Chapter 12
Online Learning: Reflections on the Effectiveness of an Undergraduate Sustainability Tourism Module

Gayle Jennings and Ulrike Kachel

Abstract Action research was used to evaluate the effectiveness of an undergraduate, tourism online module focusing on sustainability. The module emphasized "education for sustainability" along with "education about sustainability". In addition to the module, sustainability principles and practices were embedded in weekly learning engagements. Online module activities, including a learning journal, required students to use higher order thinking, which shifted their learning beyond education about sustainability to education for sustainability. The module had to be completed in order to achieve a passing grade. Several students engaged in surface learning since there was no specific grade attached to the module. Currency of links in the module required constant monitoring. Additionally, user-friendliness of the module would have been enhanced by more seamless transitions between online components and sections. The majority of students appreciated that the online module provided novelty to traditional course delivery means. Finally, the online module was determined by most students, as well as, the course convener and tutor to be an effective method to engage undergraduate tourism students in higher order thinking and student reflection regarding education about and for sustainability.

Keywords Education for sustainability • Action research • Online learning • Learning journals • Undergraduate course

G. Jennings (✉)
Imagine Consulting Group International Pty Ltd, Brisbane, QLD, Australia

Griffith University, Southport, QLD, Australia
e-mail: g.jennings@griffith.edu.au

U. Kachel
Leeds Beckett University, Leeds, UK
e-mail: u.kachel@leedsmet.ac.uk

G. Moscardo, P. Benckendorff (eds.), *Education for Sustainability in Tourism*, CSR, Sustainability, Ethics & Governance, DOI 10.1007/978-3-662-47470-9_12

12.1 Introduction

Globally, travel and tourism tertiary educators are turning their attention to values based education as a means to foster "responsible stewardship of tourism" (Liburd & Hjalager, 2010; Sheldon, Fesenmaier, Woeber, Cooper, & Antonioli, 2007). Concomitantly, there has been an increasing focus on sustainability as embedded and stand-alone curricula content in undergraduate student courses and tertiary programs of study.

This chapter draws on an action research project that developed, trialed and implemented three online modules addressing sustainability, corporate social responsibility and business ethics as an on-going "living" component of an undergraduate tourism enterprise management course. Specifically, the chapter evaluates the effectiveness of the sustainability online module from student and educator perspectives. The online learning materials included written text, visuals, audio recordings, links to related readings, video and digital media as well as self audit and personal calculators/indices related to sustainability. A key component in the module was the incorporation of a learning journal. Prompts, using questions and reflective points embedded in the module, initiated journal entries. The development of the sustainability module and its trial were informed by a critical review of extant literature associated with tertiary education contexts, "education about and for sustainability" (See Australian Research Institute in Education for Sustainability ARIES, 2008a, 2008b), the incorporation of sustainability in tertiary education programs, learning journals as educational tools, and online teaching strategies.

12.1.1 Tertiary Education Contexts

Various education philosophies inform praxis in tertiary education literature. The majority of current philosophies tend to champion the principle advocated by John Dewey (1859–1952) that: "education in order to accomplish its ends both for the individual learner and for society must be based upon experience—which is always the actual life-experience of some individual" (Dewey, 1938, p. 89). This emphasis on "experience" is further reflected in "experiential learning" approaches (Kolb, 1984); and is associated with the works of Kurt Lewin (1890–1947) as well as Jean Piaget (1896–1980), who advocated interactive learning. Other related educational philosophies include action learning (Revans, 1980) as well as approaches, which emphasize "reflexive practitioners" (Argyris, 1990; Schön, 1987). Relatedly, communities of practice, CoPs (Wenger, 1998), which are predicated on interaction and experiencing are part of tertiary education teaching philosophies. CoPs support students with regard to belonging, becoming, experiencing and doing associated with the profession into which they seek to enter (Wenger, 1998). Andragogy, that is, adult learning principles (Knowles, 1984), are also employed. Andragogy principles focus on process rather than content, adopt a problem-centered approach, and explain the "whys" of learning and evaluation of student work. Andragogy is experientially based and learning-teaching engagements acknowledge the learners self-direction and responsibility. Such self-direction and responsibility is a key component of constructivist pedagogies, in which students construct knowledge and educators facilitate learning processes (Lave & Wenger, 1991; Star & McDonald, 2007; Vygotsky, 1978).

In contrast to constructivist pedagogy are teaching philosophies predicated on the transmission of knowledge (content) modes. These modes privilege traditional teaching approaches, such as the use of lectures and case studies. In the main, however, regardless of the teaching philosophy adopted in tertiary education, lectures and case studies are still perpetuated (Pedler, Burgoyne, & Brook, 2008). Although lectures are a traditional means of curricula delivery, they have been strongly critiqued. As noted some 20 years ago, "there is an impressive body of evidence to suggest it [lecturing] is an ineffective way of realizing course aims" (Ramsden & Dodds, 1989). Earlier still it was concluded that "[t]here are times when the lecture method is an effective aid in securing desirable changes in students. There are other times when the lecture method is harmful and should be replaced by discussions or individual projects or something else" (Spence, 1928). This commentary from the 1920s, is particularly germane to the majority of students who are enrolling in contemporary western-based educational settings. This majority are part of Generation Y, a generation born in the Information Age. Generation Y are "digital natives", while other generations are considered "digital immigrants" (Prensky, 2006). Methods used to engage Generation Y in the workplace (Pendergast, 2010), have relevance to education. Ways to engage Generation Y in education include: "provid[ing] self-guided online workshops and the like", "establish[ing] a training and learning culture" and "connect[ing] theory with practice to ensure a global perspective" (Pendergast, 2010, p. 12). The use of e-journals and utilization of current communication technologies and media in learning engagements is also recommended (Pendergast, 2010). Generation Y see education as a "lifelong learn[ing]" process and part of "personal growth" (Pendergast, 2010, p. 12). Their educational needs have complementarities with the principles of adult learning.

Bearing the preceding in mind, as educators, we need to recognize that a number of traditional approaches stymie attempts by learners to engage in their own learning. As Boomer noted "I hate it when other people presume to take responsibility ... for something which I am trying to do. ... I put my mind and aspiration into neutral ..." (Boomer, 1982, p. 3). So too, is this the case for tertiary students, when educators provide prescriptive instructions for completing assessment tasks akin to applying formulae. These types of approaches reinforce surface learning with an emphasis on reproduction. Based on our critical review of extant literature, by choosing to incorporate online learning modules into the suite of learning-teaching engagements, we were attempting to develop students' communication skills, and "their ability to see relationships within what they have learned and to perceive their field of study in a broader perspective" (Ramsden, 1992, p. 20). We were also attempting "to stimulate an enquiring, analytical and creative approach, encouraging independent judgment and critical self-awareness" (Ramsden, 1992, p. 20). We were also aiming to develop critical thinking skills as well as stimulate creativity and innovation. To that end, we were attempting to foster deep learning approaches to student learning as opposed to surface learning approaches (Marton & Säljö, 1976; Ramsden, 1992). Additionally, we were intending to increase student participation and control over the learning context to reduce lecturer participation and control (Brown & Atkins, 1988).

12.1.2 Sustainability

Numerous commentators of tertiary education have noted the need for addressing sustainability in a variety of programs (ARIES, 2008b; Holmberg et al., 2008) as well as a need for education regarding stewardship in tourism studies (Liburd & Hjalager, 2010; Sheldon et al., 2007). In particular, the Australian Research Institute for Education in Sustainability, ARIES, has stated:

> [i]n response to the challenge of sustainability, there is a need for managers and leaders who have the capacity to create strategies and drive change for sustainability. Corporations and business schools have an important role to attract, develop and foster these abilities (ARIES, 2008b, p. 13).

With regard to such roles, it is important to differentiate between "education about sustainability" and "education for sustainability". The former is content and information related. The latter, that is, education for sustainability as purported by ARIES entails:

- A focus on the **future** and *capacity* to create a sustainable future—more than just focusing on trying to problem-solve our way out of our current situation
- Less emphasis on science and technology delivering solutions in isolation of the social context, and more on **participatory action**
- Less emphasis on people as the problem and more on seeing *people* as **agents of change**
- Less emphasis on awareness-raising and more emphasis on **critical reflection and systemic thinking**
- A focus on people influencing **structural change**—not just on individual environmental actions
- More focus on **changing mental models** that influence decisions and actions rather than solely changing individual attitudes (ARIES, 2008a, p. 9).

In our online modules for sustainability, we focused on education about sustainability; however, our primary focus was on education for sustainability. In the introduction to the module students were informed that:

> The purpose of this module is to provide you with an introductory overview to educate you "about and for sustainability". The primary focus is sustainability with regard to your professional and practitioner work-related career paths. From time to time, however, you will be challenged to consider sustainability with regard to day-to-day living and the sustainability of your current practices (Griffith University, 2008, p. 1).

The online module covered the following topics: What is Sustainability?; Sustainability and you; Towards Sustainable Praxis; Sustainability indices, measures and tools; The University and sustainability; Making sustainability part of your praxis; About Sustainability—Links to ethics and corporate social responsibility. Key emphasis in the module was to engage students in "Reflecting on their future role with regard to sustainability within their life and working career paths". This was achieved through the use of critical questions and critical student reflections using a learning journal. Additionally throughout the weekly lectures with guest speaker inputs sustainability was positioned as part of professional praxis. That is to say sustainability had to become an integral part of teaching rather than an add-on component (Holmberg et al., 2008; Perdan, Azapagic, & Clift, 2000). Consequently, sustainability was also related to all assessment tasks and a sustainability criterion was part of assessment criteria for each assessment task.

Learning journal entries were an important component of the online learning experience because of their links to critical thinking and reflection. "Reflection, and allowing students time to reflect upon what they have learnt is important. The process of reflection, reviewing and making sense out of what has been done is essential if real learning is to take place" (Reid, 1988, p. 132).

The questions used in the sustainability online module to trigger journal entries were based on Bloom's revised taxonomy of questioning because of its original development for academic education usage (Bloom & Krathwohl, 1956). The revised questions have been categorized as: questions, which involve remembering, understanding, applying, analyzing, evaluating and creating (Anderson & Krathwohl, 2001). The learning journal questions were predominantly higher order questions involving analyzing, evaluating and creating. The following are examples taken from the sustainability module.

- **Analyzing**: What is your opinion—is one pillar more important than the rest? Justify your answer. In your opinion, what are the linkages between corporate sustainability and corporate social responsibility?
- **Evaluating**: (1) What is your opinion about the measures noted for measuring sustainability? (2) If you were to use a sustainability index for your profession/project which would you use and why?
- **Creating**: (1) Can you design something better? (2) If you were to prepare a framework to evaluate sustainability for your profession/project what would it entail?

12.1.3 Online Educational Technologies

While technology is increasingly being used in tertiary education (Ellis, Goodyear, Prosser, & O'Hara, 2006; Knight, 2010), evaluations of its effectiveness to facilitate student learning remain varied (Turney, Robinson, Lee, & Soutar, 2009). On one hand, it is lauded for increasing student engagements with learning (Maier & Warren, 2000; Sanders, 2006). On the other hand, it has been noted for its encouragement of surface learning by students (Saunders & Klemming, 2003). The use of online technology is also varied, ranging from use as a repository for course administrative and related teaching materials to a replacement for real time/synchronous face-to-face traditional learning-teaching engagements (Turney et al., 2009). Despite the uncertain nature of online educational technologies' effectiveness and uses, one of the reasons for using e-learning is to improve the quality of learning (Bates, 1997). Another reason is to expand the repertoire of learning experiences for students (Star & McDonald, 2007). These two reasons influenced the use of online learning in our action research project. We did not choose to use a complete integrated online learning experience for students rather we used a 'blended learning' approach (Ellis et al., 2006; Knight, 2010). We did this because we recognized that learners in online environments may lack a social place to interact with others regarding their learning. This is a critical element for scaffolding learning (Vygotsky, 1978). Overall, we aimed to provide a community of practice by offering a variety of learning teaching engagements, which matched our emphasis on social learning theories informed by our constructivist pedagogy. Communities of practice differ from learning communities, that is, "bounded group [s] of students involved in cooperative learning online" (Misanchuk & Anderson, 2001). These in turn differ from other online communities, which are formed based on a common interest, specifically, a community of interest (Misanchuk & Anderson, 2001).

12.2 Methodology

The approach used to determine the effectiveness of the undergraduate tourism modules, including the module that focused on sustainability, was action research, particularly classroom action research. Classroom action research was chosen since it is an effective way to implement, monitor and evaluate changes to educational curricula (Jennings, Scantlebury, & Wolfe, 2009). As an approach, the roots of action research are in social psychology and it is founded on the work of Kurt Lewin (1946). Action research is now used across differing disciplines and fields of study to facilitate positive change or improvement to praxis (Kemmis & McTaggart, 2005). One interpretation of action research notes that it aims "to identify problematic situations or issues considered by participants to be worthy of investigation in order to bring about critically informed changes to practice" (Cornwell, 1999). As a research approach, it is a cyclical process of reflecting, planning, acting and observing (Kemmis & McTaggart, 2005). Paradigmatically, action research has an ontological or worldview that is collectively constructed, and an epistemology which is embodied, situated and hermeneutically framed (Jennings, 2010). In this study, the methodology was qualitatively founded. Axiologically, the study was value leaden with a transformational intent of improving educational curricula, specifically, values related to sustainability and, in the other modules, corporate social responsibility and business ethics.

The action research participants were the course convener, tutor, and tourism students involved in the tourism enterprise management course. Empirical materials included qualitative conversations between convener and tutor, convener or tutor with students, written learning-teaching materials generated by students as well as learning journals kept by the convener and tutor. The empirical materials were interpreted using constant comparison. In reporting the outcomes of our interpretations, we have clustered student responses into themes as well as to condense student commentaries into emblematic re-constructions to maintain student anonymity.

12.3 Findings and Discussion

During the semester, students engaged in the modules in lieu of regular face-to-face learning-teaching interactions. The students indicated that the modules were a change from face to face interaction and added variety to the course's suite of learning contexts. The independent study experiences increased student reflections and there was a strong appreciation of having a learning-teaching engagement, which did not have to be supported by academic references but rather with student critical reflections.

While on the surface, it may appear that the modules provided students with stand alone learning opportunities, as already noted this was not the case. Students were regularly required to engage with the modules' content in lectures, tutorials and assessment tasks. Student reflections on their own learning indicated that they had gained broader perspectives and understandings from engagement in the modules and used those in other learning components of the course. Reflections on the effectiveness of the online modules, in this chapter, education about and for sustainability and online-related learning journals are now presented.

12.3.1 Effectiveness of Online Educational Technologies

Students' feedback on the effectiveness of the online module clustered around perspectives associated with learning styles and usability. With regard to learning styles emblematic themes were: "not how I learn"; "suits my learning style", "change from lectures", "appreciated self directed learning", "not a self directed learner", and "provided a positive challenge". The usability of the module received differing comments: "easy to understand", "the sustainability module was time consuming", "problems accessing links", "multimedia inputs facilitated interest and variety", "perceptions of repetition in subsections", "an enjoyable experience" and a "frustrating experience".

12.3.1.1 Areas for Improvement

Rather than opening a new page for each sub-section and then having to close it and return to the home page again to get to the next sub-section, a student suggested that the modules should open and close sub-module sections from within the home page. Another comment made specifically regarding the sustainability module related to the range of video-streams provided so students could self-select based on interest. Some student feedback indicated that there were too many video clips and these should be reduced in number. Future use of the modules, should involve coaching of students so they may make effective decisions when offered diversity and choice in media with which to engage. Further, clarification regarding perceptions of repetition of content or themes needs to be undertaken. If relevant, commentary regarding concept overlaps in the modules resulting from disparate and integrated "treatment" should be made. Additionally, review of the time taken to complete modules needs to be re-evaluated due to varying student backgrounds and English language competencies.

12.3.2 Effectiveness of Education About and for Sustainability

In this section, we reflect on education about sustainability, education for sustainability as well as students' reflections on personal praxis. Here our emphasis is on what students think about these terms and what they learnt or need to learn.

12.3.2.1 Education About Sustainability

In reflecting on what sustainability was prior to engaging with the module, the majority of students' definitions were predicated on resource usage for current and future generations. A number of students commenced the module with an environmental emphasis in their definitions of sustainability, by the end of the module; an understanding of the full facets of sustainability had been obtained. Additionally, the students reflected that the Brundlandt Report definition World Commission on Environment and Development, WCED (1987) was still relevant for today's world. Several students expressed disappointment with how little had been achieved since the report's publication.

12.3.2.2 Education for Sustainability

Students were asked to reflect on what 'sustainability' means for them as a future professional and/or practitioner. Students' reflections were numerous and encompassed perspectives on personal roles and responsibilities; collective responsibilities; the importance of sustainability for industry, businesses, and destinations, and the influences of consumer expectations on businesses with regard to sustainability practices. The following reconstructed comment is emblematic of students' future praxis: “As a professional, in a tourism management role, I expect to take into consideration and apply the major elements of sustainability frequently.”

12.3.2.3 Students Reflections on Personal Praxis

Students were asked to self-evaluate their current personal sustainability practices. Several students considered themselves as already being quite sustainable in their daily practices, making comments similar to: “Sustainability is part and parcel of my everyday life and I always attempt to live effectively in order to sustain the future.” The majority of students, however, acknowledged that they could do better and nominated practices, which they could improve upon. One student exclaimed: “I was surprised with my green house emissions calculation. While I try to live sustainably, I still have room for improvement like recycling and using buses more!” Throughout students’ reflections on their own practices, they stated the importance of sustainability. An emblematic reconstructed comment demonstrates this: “If we want to achieve sustainability, then everyone has a role to play.” One student also reflected that instead of sustainability that there were more urgent issues that needed to be addressed, such as, famine and poverty.

12.3.3 Effectiveness of Online-Related Learning Journals

The learning journals as a means to develop critical thinking and reflection were very effective. Students indicated that the opportunity to make a learning journal entry at the end of activities or sub-sections both provided space for critical thinking as well as facilitated further reflection on topics.

12.3.3.1 Areas for Improvement

The online modules used the same icon to identify learning journal entries and learning activities. Each time the icon was used it was accompanied by written text indicating if it was a ‘Learning Journal’ or an ‘Activity’. This generated confusion for some students not knowing when they should make an entry into their journal or perceiving that the journal entries were not connected or cohesive and difficult to monitor with regard to what had been completed and what had not. Future maintenance of the modules needs to distinguish icons and increase ease of tracking journal entries.

12.3.4 Overall Educator Reflections

We aimed to develop deep rather than surface learning in the students. In the main, the majority of students engaged in deep learning when engaging with the online modules and completing the learning journal tasks. Several students, however, did not and only superficially engaged with the modules and the journals tasks. Students were aware that the online learning journals were being read rather than assessed by the course convener (Marton & Säljö, 1976). Students also understood that the journals were required in order to be considered for a passing grade in the course. Subsequently, those students, who chose to engage as surface learners, made entries accordingly, since there were no marks to be allocated for completing the task. These students demonstrate an example of the hidden curriculum (Snyder, 1971). The students responded to the curriculum (the online learning modules) based on their perception of its importance in relation to assessment (no marks—submission only) (Star & McDonald, 2007). The majority of students, however, engaged as deep learners reflectively considering connections between ideas and concepts as well as previous learnings. These students were focused on making meaning as they engaged with the online learning teaching modules.

12.4 Conclusions

Improvements need to be made to the "living" content of all online modules. Specifically, there is a constant need for maintenance and review of currency of materials. Technology must be reliable (Alexander, 2010), and support to educators and for students, the endpoint users, must be maintained. If the technology does not work then students will turn away from the learning environment (Alexander, 2010). Several of the links in our modules became broken in the course of the semester. Recognizing this, we provided complementary materials with other links elsewhere in the course's online environment. Some students did not check these complementary support materials for other links. Additionally, the learning journal was unable to be entered on line and had to be completed using a concurrent open Word document. Since our research was conducted, software packages are now being used, which enable journal entries to be entered online. This facilitates greater connectivity between the module and student learning.

Overall, the sustainability module along with the two other modules on corporate social responsibility and business ethics were an innovation to the course's learning-teaching engagements and were viewed by the students as effective means to reflectively engage with course content. That being said, several students, adopted surface learning approaches wherein the students did not engage in the tasks but rather completed them superficially as they did not value the online learning engagement. The degree to which student choices with regard to surface versus deep learning remains as always the responsibility of the student. However, it is our responsibility to contextualize learning experiences in order for students to see the worth and value of these experiences to their education and to their future.

On-going maintenance and re-design is a constant in any online learning module development in order to ensure currency of content and state of the art-ness of the online tool itself. With regard to the focus of this chapter, sustainability, this is critical to enhance student learning about and more importantly for sustainability.

References

Alexander, S. (2010). E-learning developments and experiences. *Education + Training, 43*(4/5), 240–248.

Anderson, L. W., & Krathwohl, D. R. (Eds.). (2001). *A taxonomy for learning, teaching and assessing: A revision of Bloom's taxonomy of educational objectives: Complete edition.* New York: Longman.

Argyris, C. (1990). *Overcoming organizational defenses: Facilitating organizational learning.* Boston, MA: Allyn and Bacon.

Australian Research Institute in Education for Sustainability ARIES. (2008a). *Partnering business schools and corporations: Education about and for sustainability in Australian business schools: Stage 3 [Contract Document].* Sydney: ARIES.

Australian Research Institute in Education for Sustainability ARIES. (2008b). *Definitions: A working document* (p. 9). Sydney: ARIES.

Bates, A. W. (1997, June 18–20). *Restructuring the university for technological change.* Paper presented at *What Kind of University?*, The Carnegie Foundation for the Advancement of Teaching, London. http://www.educause.edu/Resources/RestructuringtheUniversityforT/151339. Accessed 19 May 2010.

Bloom, B. S., & Krathwohl, D. R. (1956). *Taxonomy of educational objectives: The classification of educational goals, by a committee of college and university examiners. Handbook 1: Cognitive domain.* New York: Longmans.

Boomer, G. (1982). Turning on the learning power: Introductory notes. In G. Boomer (Ed.), *Negotiating the curriculum: At teacher-student partnership* (pp. 1–7). Sydney: Ashton Scholastic. See p. 3.

Brown, G., & Atkins, M. (1988). *Effective teaching in higher education.* London: Routledge.

Cornwell, S. (1999). An interview with Anne Burns and Graham Crookes. *The Language Teacher, 23*(12), 7–9. http://www.jalt-publications.org/tlt/articles/1999/12/cornwell. Accessed 18 May 2010.

Dewey, J. (1938). *Experience and education* (p. 89). New York: Collier Books.

Ellis, R. A., Goodyear, P., Prosser, M., & O'Hara, A. (2006). How and what university students learn through online and face-to-face discussion: Conceptions, intentions and approaches. *Journal of Computer Assisted Learning, 22*, 244–256.

Griffith University. (2008). *3202HSL sustainability module* (p. 1). Gold Coast, QLD: Griffith University.

Holmberg, J., Svanström, M., Peet, D.-J., Mulder, K., Ferrer-Balas, D., & Segalàs, J. (2008). Embedding sustainability in higher education through interaction with lecturers: Case studies from three European technical universities. *European Journal of Engineering Education, 33* (3), 271–282.

Jennings, G. (2010). *Tourism research* (2nd ed.). Brisbane: Wiley.

Jennings, G., Scantlebury, M., & Wolfe, K. (2009). Tertiary travel and tourism education: Using action research cycles to provide information on pedagogical applications associated with reflexivity, team-based learning and communities of practice. *Journal of Teaching in Travel & Tourism, 9*(03–04), 193–215.

Kemmis, S., & McTaggart, R. (2005). Participatory action research: Communicative action and the public sphere. In N. K. Denzin & Y. S. Lincoln (Eds.), *The sage handbook of qualitative research* (3rd ed., pp. 559–603). Thousand Oaks, CA: Sage.

Knight, J. (2010). Distinguishing the learning approaches adopted by undergraduates in their use of online resources. *Active Learning in Higher Education, 10*(1), 67–76.

Knowles, M. (1984). *Andragogy in action* (1st ed.). San Francisco: Jossey-Bass.

Kolb, D. A. (1984). *Experiential learning: Experiences as a source of learning and development*. Englewood Cliffs, NJ: Prentice-Hall.

Lave, J., & Wenger, E. (1991). *Situated learning: Legitimate peripheral participation*. Cambridge, UK: Cambridge University Press.

Lewin, K. (1946). Action research and minority problems. *Journal of Social Issues, 2*, 34–46.

Liburd, J., & Hjalager, A.-M. (2010). Changing approaches towards open education, innovation and research in tourism. *Journal of Hospitality and Tourism Management, 17*, 12–20.

Maier, P., & Warren, A. (2000). *Integrating technology in learning and teaching*. London: Kogan Page.

Marton, F., & Säljö, R. (1976). On qualitative differences in learning. II—Outcome as a function of the learner's conception of the task. *British Journal of Educational Psychology, 46*, 115–127.

Misanchuk, M., & Anderson, T. (2001, April 8–10). Building community in an online learning environment: Communication, cooperation and collaboration. In *Proceedings of the Annual Mid-South Instructional Technology Conference*, Murfreesboro, TN. See page 2.

Pedler, M., Burgoyne, J., & Brook, C. (2008). What has action learning learned to become? *Action Learning: Research and Practice, 2*(1), 49–68.

Pendergast, D. (2010). Getting to know the Y generation. In D. Pendergast, P. Beckendorff, & G. Moscardo (Eds.), *Tourism and generation Y* (pp. 1–15). Wallingford, Oxfordshire: CABI.

Perdan, S., Azapagic, A., & Clift, R. (2000). Teaching sustainable development to engineering students. *International Journal of Sustainability in Higher Education, 1*(3), 267–279.

Prensky, M. (2006). Listen to the natives. *Educational Leadership, 63*(4), 8–13.

Ramsden, P. (1992). *Learning to teach in higher education*. London: Routledge.

Ramsden, P., & Dodds, A. (1989). *Improving teaching and courses: A guide to evaluation*. Parkville, VIC: The Centre for the Study of Higher Education, The University of Melbourne. See page 36.

Reid, J. (1988). Negotiating education, appendix D. In S. Kemmis & R. McTaggart (Eds.), *The action research planner* (3rd ed., pp. 111–132). Melbourne: Deakin University Press. See page 132.

Revans, R. (1980). *Action learning: New techniques for management*. London: Blond & Briggs.

Sanders, R. (2006). The "imponderable Bloom": Reconsidering the role of technology in education. *Innovate, 2*(6). http://innovateonline.info/pdf/vol2_issue6/The_Imponderable_Bloom_Reconsidering_the_Role_of_Technology_in_Education.pdf. Accessed 18 May 2010.

Saunders, G., & Klemming, F. (2003). Integrating technology into a traditional learning environment: Reasons for and risks of success. *Active Learning in Higher Education, 4*(1), 74–86.

Schön, D. A. (1987). *Educating the reflective practitioner: Toward a new design for teaching and learning the professions*. San Francisco: Jossey-Bass.

Sheldon, P., Fesenmaier, D., Woeber, K., Cooper, C., & Antonioli, M. (2007). Tourism education futures: 2010–2030: Building the capacity to lead. *Journal of Teaching in Travel and Tourism, 7*(3), 61–68. See page 63.

Snyder, B. R. (1971). *The hidden curriculum*. Cambridge, MA: MIT Press.
Spence, R. B. (1928). Lecture and class discussion in teaching educational psychology. *Journal of Educational Psychology, 19*, 454–462. See page 462.
Star, C., & McDonald, J. (2007). Embedding successful pedagogical practices: Assessment strategies for a large, diverse, first year student cohort. *International Journal of Pedagogies and Learning, 3*(2), 18–30.
Turney, C. S. M., Robinson, D., Lee, M., & Soutar, A. (2009). Using technology to direct learning in higher education: The way forward? *Active Learning in Higher Education, 10*(1), 71–83.
Vygotsky, L. S. (1978). *Mind in society: The development of higher psychological processes* (M. M. Lopez-Mrillas Cole, A. R. Luria, & J. Wertsch, Trans.). Cambridge, MA: Harvard.
Wenger, E. (1998). *Communities of practice: Learning, meaning, and identity*. Cambridge, UK: Cambridge University Press.
World Commission on Environment and Development, WCED. (1987). *Our common future*. Oxford: Oxford University Press.

Snyder, B. R. (1971). *The hidden curriculum*. Cambridge, MA: MIT Press.
Spence, L. J. (1924). Lecture and class discussion in teaching educational psychology. *Journal of Educational Psychology*, 15, 454–462. See page 48.

Chapter 13
Practice What You Teach: Teaching Sustainable Tourism Through a Critically Reflexive Approach

Erica Wilson

Abstract The use of critical theory, while gaining momentum in some areas of tourism research, is rarely extended to the teaching of tourism in higher education. This paper explores the role and value of critical pedagogy as it relates to how we teach tourism studies. In particular, the paper adopts a reflexive, autobiographical approach to outline my own efforts to build Critically Reflective Practice (CRP) into an undergraduate sustainable tourism class. In doing so, I aim to address concerns that there is little guidance on how to 'do' critically reflective practice. As such, practical examples are provided to assist tourism academics who are interested in incorporating CRP into their curricula. Ultimately, it is argued that CRP has the potential to transform the teaching experience, (re)motivate the lecturer's enthusiasm for the teaching role, and provide a stronger nexus between teaching and research.

Keywords Critical pedagogy • Critically reflective practice • Sustainable tourism • Higher education

13.1 Introduction

Alongside the emergence of sustainable development, sustainable tourism has been a key area of study now for almost four decades. This research agenda has focused on alternatives to mass tourism, as well as addressing the 'triple bottom line' of tourism's socio-cultural, economic and environmental impacts (Weaver, 2005). While there has been a concomitant increase in sustainable/eco-tourism focused subjects (and even whole degrees in some cases), the literature surrounding the pedagogy and teaching of sustainable tourism appears rare (Deale & Barber, 2012; Wilson & von der Heidt, 2013). As we near the end of the United Nations Decade of Education for Sustainable Development (2005–2014), and as this very book explores, many questions remain about how we educate for sustainability in tourism higher education.

E. Wilson (✉)
Southern Cross University, Lismore, NSW, Australia
e-mail: erica.wilson@scu.edu.au

G. Moscardo, P. Benckendorff (eds.), *Education for Sustainability in Tourism*, CSR, Sustainability, Ethics & Governance, DOI 10.1007/978-3-662-47470-9_13

This chapter discusses how critically reflexive practice (CRP), as part of a critical theory pedagogy, can be used within the teaching of sustainable tourism in the university context. 'Critical thinking' and 'critical reflection' are highly valued concepts in university education today (Biggs & Tang, 2007; Brookfield, 1995; Fisher, 2003). As Gabennesch (2003, p. 36) pragmatically puts it: 'nearly everyone is in favour of critical thinking'. We all want our tourism students to be critical thinkers, but what does being 'critical' *really* mean? Further, there is little practical guidance about how we translate critical pedagogy into the tourism 'classroom', particularly when many tourism degrees are couched in a business/management paradigm (Belhassen & Caton, 2011; Wilson & von der Heidt, 2013).

To draw out these discussions, this chapter will draw on my own reflexive experiences in teaching sustainable tourism, and demonstrate tangible examples that can be used to incorporate CRP in tourism teaching. '*Sustainable Tourism*' is an elective, second-year unit—offered both internally and externally via distance education—within the Bachelor of Business in International Tourism Management at Southern Cross University in Australia. In this class, students are introduced to the debates, theories and definitions surrounding the concepts of sustainable development and sustainable tourism. Until recently, it was also a core unit for students enrolled in the Bachelor of Environmental Tourism Management (no longer offered). The stated aim of Sustainable Tourism engages upfront with critical pedagogy, and is as follows:

> This unit critically examines the concept of 'sustainable development', and how it applies to the planning and management of tourism. Using sustainability as a basis, students are introduced to the social, cultural and environmental impacts of tourism, both in Australia and internationally. Students are engaged in the processes of critical thinking and reflection throughout the unit and its assessment.

The syllabus in the first half of *Sustainable Tourism* is dedicated to understanding, deconstructing and critiquing the concept of 'sustainable development' as it applies to tourism. Students are familiarised in the history of the modern environmental movement, documenting how alternative paradigms have challenged traditional, anthropocentric approaches. Once this socio-ecological basis is established, the unit turns its attention to sustainable tourism, and its emergence via the cautionary platform of tourism research (Jafari, 1990). The second half of the unit is dedicated to the practical issues of tourism and recreational planning, particularly in relation to the natural environment and ecotourism. These issues are also grounded in a critical pedagogical framework.

The stated objectives of the unit are that, on completion of this unit, students should be able to: (1) analyse the concept of sustainable development, and how it applies to tourism; (2) critically assess the impacts of tourism locally, nationally and internationally; (3) recognise the core elements central to nature-based tourism and ecotourism; (4) define best-practice principles for a sustainable tourism industry; and, (5) practise being critical and reflective thinkers, who can examine their values, beliefs and assumptions.

The unit is made up of two key assessment items: a Critical Skills Blog (40 %), and a Critical Essay (60 %). I deliberately removed the exam from *Sustainable Tourism* class a number of years ago, as I no longer felt it was the most effective way of assessing students' learning and reflection around the area of sustainability in tourism. Similar points have been made by other scholars of critical theory (Salazar, 2013).

13.2 Theoretical/Conceptual Framework

Grounded firmly in the ideals of critical theory, and particularly the works of humanist education scholar Paolo Freire (1970), CRP embraces a transformational, emancipatory and often radical learning agenda, transgressing common understandings of 'critical thinking' (Biggs & Tang, 2007; Hooks, 2003). CRP is not only concerned with critique in the sense of getting students to see different sides of an argument. Those working from a critical approach also seek to challenge dominant hegemonies, and call for societal and ecological change through a more humanistic pedagogy (Brookfield, 2005; Fullagar & Wilson, 2012; Salazar, 2013; Wals & Jickling, 2002).

For these reasons, teaching university students critical, liberal and humanistic thinking skills is important to critical scholars (Cuncliffe, 2004; Fulop, 2002; Holmes et al., 2005). Without these skills, it is also arguable whether students can survive effectively in the workplace and in the wider communities in which they live. Belhassen and Caton (2011, p. 1395), writing on the need for critical pedagogy in tourism education, argue that:

> ...academia would be better serving [students] with educational preparation that cultivates more critical understandings of social systems, such that students reconcile themselves to the limits of their own power ... Such a view restores human agency, including moral agency, to those who seek to work as coordinators and leaders in the tourism industry.

Moreover, Fulop (2002) thinks it is not enough to say that students must think critically; we as teachers must 'practice what we preach' (Kearins & Springett, 2003; Welsh & Murray, 2003). Based on a critical pedagogy, it is important for both teachers and students to learn to be 'reflective practitioners' (Schön, 1983). As Merriam, Caffarella, and Baumgartner (2006) suggest, reflective practice requires a 'deliberate slowing down to consider multiple perspectives [while] maintaining an open perspective' (p. 173).

Sustainable tourism is a useful vehicle for considering such multiple perspectives, given that 'sustainability' itself has stemmed from an alternative response to the capitalist ideology of the Dominant Western Environmental Paradigm (DWEP) (Hunter, 2002; Weaver, 2005). A critical approach to sustainable tourism rests largely upon questioning of the DWEP, and recognition that tourism is a political and socially-constructed phenomenon, in which some voices and agendas are heard—and others are not. Sustainable tourism education emphasises the *ecological* dimension where there is synergy among all aspects of the educational process (Sterling, 2010). Moreover, sustainable development and sustainable tourism are some of the most contested ideas in the academic literature today (Atkinson, Dietz, & Neumayer, 2007; Jamal, 2004), and thus ripe for ongoing critique and deconstruction. There is no better place to start with this critique than with our students.

13.3 Description of the Teaching Activities

It would be too simplistic to describe CRP through a single teaching activity, or even a set of activities. CRP is an overarching philosophical and values-based framework, underpinning everything from the development of one's teaching philosophy, through to study guides and syllabi, teaching and learning activities, as well as the assessment and evaluation of student work. While CRP does prescribe an ideological framework for pedagogy, it is also lived and acted out in the everyday 'classroom', with real teachers, real students and a living curriculum (Cockburn-Wootten, Henderson, & Rix, 2005; Kearins & Springett, 2003).

Critically reflective practice in teaching is usually based on three core, interrelated, principles (Holmes, Cockburn-Wootten, Motion, Zorn, & Roper, 2005):

1. *Critique*: there must be critique and deconstruction of our cultures and of the dominant culture;
2. *Questioning of Assumptions*: a recognition that knowledge(s) are socially constructed and competing, and based on a set of assumptions about the world;
3. *Praxis*: teachers must draw from students' own lived experiences, and in critically reflecting on these draw a firm link between theory and practice.

Using these three 'cornerstones' of CRP, I will now demonstrate, using examples, how they are enacted in my own teaching activities and practices in *Sustainable Tourism.*

13.4 Building Critique

From the outset (indeed, from the time the students read their unit information guide, as outlined above in the unit aim and objectives), students are aware that *Sustainable Tourism* is grounded in critical theory, and that they will learn to be critically reflexive practitioners—even though they may not be aware yet what that will mean. Yet the concepts are not meant to be employed in an elitist or non-transparent way, or in an effort to confuse students. As an educator, I understand that to start this journey means I must first engage with students in a process of learning about what critical thinking and critical reflection mean. In the first lecture, I use two slides (see Fig. 13.1) to outline the difference between 'critical thinking' and 'critical reflection'.

Fig. 13.1 Outlining critical thinking and critical reflection

CRITICAL THINKING

'Detecting the assumptions underlying another's position or text' (Fisher 2004, p 1)

- Commonly expected and taught in higher education
- Reasoned judgement
- Logical, abstract reasoning
- Skepticism, questioning
- Distinguish fact from opinion

CRITICAL REFLECTION

'A process attached to looking at one's own positioning: requires self-reflection and self-awareness' (Fisher 2004, p1)

- Going beyond the what and the how to the WHY
- Always asking *cui bono: who benefits?*
- Recognising the role of power and hegemony
- Located within wider political, social and cultural forces
- Conscious and self-aware
- *May* lead to change and action

To put some boundaries around this concept, I also highlight what critical thinking is *not*. That is: it is *not* some form of psychotherapy that I am using in class, where students are required to reveal deep, personal sharing with which may lead them to feel vulnerable. It is also not about harshly 'criticising' others, or being the loudest or most talkative in the class. I reiterate, again in the first lecture, that CRP is not an outcome or end that can be easily measured or assessed: *it is the PROCESS that we are interested in; or a journey of stopping, reflecting, and learning.*

13.5 Questioning Our Assumptions

Another hallmark of CRP is learning to question one's values, beliefs and assumptions—or one's ontological way of looking at the (tourism) world around them. The students often ask why is this important. My response is that by recognising that actions and policies in sustainable tourism are based on assumptions, we will be in a better position to understand them and be open to change as new facts and new contexts emerge. As an example: we may have to challenge assumptions like 'mass tourism is good', 'ecotourism is good', or that 'sustainable tourism is the answer to all our problems' or that 'poor people can be saved by tourism'. I use a one page handout from Stephen Brookfield (2006), who has written widely on critically reflexive practice in adult education. As part of this handout, Brookfield (2006, no page) says that:

> [Our] assumptions are sometimes correct. At other times, however, the assumptions we base our decisions on have never been examined . . . To make good decisions in life we need to be sure that these assumptions are accurate and valid—that they fit the situations and decisions we are facing. . . . Critical thinking describes the process we use to uncover and check our assumptions. First we need to find out what our assumptions are.

Fisher (2004) differentiates clearly between values, beliefs and assumptions: *Values* refer to what we think is worthwhile and reflect how we feel the world should be. The central question in outlining our values is *what is important to me? Beliefs* reflect our feelings of how the world is, rather than how it *should* be. The key question here is *what do I think is true? Assumptions* are related to our values and beliefs, but they are presuppositions of how the world operates. Fisher suggests that assumptions are in fact a combination of unquestioned values plus unquestioned beliefs.

To aid students in thinking about their own values, beliefs and assumptions and democratise the classroom, I outline my own on powerpoint slides in the first lecture of the class (see Fig. 13.2), as well as some commonly held statements they may hear around them.

ERICA'S VALUES

- What is important to me is that humans respect themselves, each other and the natural world
- I value fairness, social inclusivity, and transparent decision-making, responsibility for self, and the power of community relationships
- At the personal level, I value self-reflection, humour and the ability to see others' points of view

ERICA'S BELIEFS

- I believe that travel can be good for people.
- It's true that children shouldn't watch too much TV
- I believe that climate change exists, is caused by humans and that the natural environment is in decline
- I believe that critical thinking is good, because it makes us challenge our beliefs

ERICA'S ASSUMPTIONS

- I assume that education and travel are good for people
- I assume that questioning one's assumptions is a healthy thing
- I assume that students will be interested in this unit, because they have enrolled in it!

Fig. 13.2 Role of the teacher's values, beliefs and assumptions

This was a process I had not done before, and was a little reticent in exploring with students. I also discuss my own struggles and challenges in trying to live a 'sustainable' rural life, and building our own energy-efficient home. My values, beliefs and assumptions are written out clearly as simple statements, so that students can see the differences between values, beliefs and assumptions, and how they might be worded. Putting my own assumptions forward helps to demonstrate my own reflexivity and transparency; I am trying to tell students: 'this is who I am, why I teach the way I do, and why sustainable tourism is important to me'.

This is then followed by an activity whereby students are broken into small groups and asked to reflect upon their own beliefs, values and assumptions in relation to tourism and the environment. This had the effect of 'breaking the ice' in the first lecture of the semester, as well as grabbing students' attention and avoiding a passive learning experience. External students could access the lecture notes, and were required to participate in moderated on-line discussions, about their values, beliefs and assumptions. These thinking skills were then honed through a critical thinking assessment, part of which is described below.

13.6 Praxis: Linking Theory to Practice

The idea of praxis is that there is a merging between practice and theory. On reflection, I had become too focused on *telling* students about the theories in sustainable development/tourism, without perhaps thinking about how this related to their own positions and life experiences. Thus I made a deliberate choice, grounded in and supported by critical pedagogy, to actively welcome the first person voice in the first assessment, and to allow students to try to reflect on their own values, beliefs and assumptions. To do this, an assessment item called a 'Critical Skills Blog', worth 40 %, was created. The aim of this assignment, which consisted of two online questions, as outlined to students,

> ....is to develop and encourage your *critical thinking and reflection* on sustainable tourism issues. These assignments are designed to help you think critically about the assumptions you have, as well as those that others hold, with regard to tourism and the environment. This is why the public face of the blog is central to this assessment; students can read and consider others' viewpoints, and learn from them.

The first Blog assignment was titled *Critical reflection: Understanding your values, beliefs and assumptions*. For Blog 1, students were asked to read Fisher's (2004) paper titled '*Critical Reflection: What is it and How Do You Do It?*', the first paper in their Book of Readings, again demonstrating how critical pedagogy and critical thinking are front and centre, and a platform used to guide the entire subject. Fisher's article is an easily accessible paper that helps students understand critical theory, and how it is practiced via understanding and uncovering one's values, beliefs and assumptions. For their assignment, students are asked to first describe, after reading Fisher's paper, what they understand by 'critical reflection'. They are then asked to outline one value, one belief and one assumption, and why they might have each of these. They are directed to, if at all possible, relate these examples to the natural environment, tourism or sustainability somehow, but if not, students can draw more widely on values they may have generally. It can be a challenging task, but brings students' attention again to why critical thinking and reflection are important in learning about sustainable tourism.

Another in-class technique I use to demonstrate praxis, borrowing from Kearins and Springett's (2003), is to get students thinking about sustainable development theory in the context of their own lives. This connects directly with our Week 4 syllabus topic on 'Sustainable Tourism', and where students read Hunter's (2002) theoretical chapter on the strong to weak sustainability spectrum. After a lecture on this concept, I draw an imaginary spectrum or line across the classroom floor. One end of the spectrum is identified as 'very strong' sustainability, and the other 'very weak'. I ask the students to try, as best they can, to place themselves along this spectrum in terms of their own environmental action, and make the point that we are not involved in judgement about who is the 'greenest'. Students then stand at different points along the line, and volunteers are asked to explain why they have put themselves there. I do the same. This activity works well to demonstrate the link between theory (the ideological spectrum) and practice (the students' lives/contexts). The physical movement required gets students out their seats and visibly enlivens the classroom and opens up room for informal discussion. Of course, it is not as easy to translate this in the distance education environment, and these are some of the challenges that need to be explored further when evaluating CRP in a rapidly changing university environment.

Another activity that works well is a 'Sustainable Development Definition Matching Game'. Here, I collect different definitions of sustainable development, from different political parties, interest groups, environmental groups and so on. I print four SD definitions out on poster size papers, stick them to the board, and then show the students, randomly, each of the four interest groups. Select volunteers are then asked to stick the matching interest group to the matching definition, and we see if they are correct. This activity helps to demonstrate the link between different SD paradigms/theories, and the reality of how these are played out in 'real life'. It is often an illuminating and enjoyable in-class technique.

Other efforts at praxis are field trips, or experiential learning. For many years, I conducted a field trip for all *Sustainable Tourism* students. We visited local tourism operators, villages and environmental agencies, and camped overnight at a community Permaculture farm across from where I live (the farm is no longer in operation so this has become difficult). The Farm was run entirely on solar power, and the students' meals were all made from the on-site garden. Students were introduced (many for the first time) to compost toilets, and how a community of permaculture students lived together in an energy-efficient, sustainable manner. Again, field trips are an excellent way in which to bring home to students how the theory of sustainable development is acted out in practice.

13.7 Evaluation of the Activities

This chapter has relied on my own reflexive accounts of embedding CRP into the teaching of sustainable tourism. It should be noted that students' reactions to the CRP approach have not yet been measured or described, apart from formal semester student feedback, where many positive comments have been received. Some students, however, reported not understanding why there was such a 'critical' focus on the unit, and could not clearly see the links to their business tourism degree. This means I have had to reflect on the techniques I have used, be clear about my intentions, and what CRP's relevance is to them and to *Sustainable Tourism*.

Generally and anecdotally speaking, however, I found I enjoyed teaching the unit more (after 8 years of the same class), and in some cases students seemed more engaged. Many of the activities described above helped to make the classroom a bit more 'fun', and reduce the power distance between myself and my students. In 2007, after implementing CRP for this first time into *Sustainable Tourism,* I received the highest student feedback score for this unit since my teaching in it began in 2003. This reflects perhaps the power of critically reflective practice (Brookfield, 2005) and my own renewed motivation to the scholarship of teaching. Indeed, other studies have demonstrated how CRP can benefit the classroom, engendering more innovative, democratic and enlightened teaching and learning environments (Cockburn-Wootten et al., 2005; Kearins & Springett, 2003).

For me, CRP provided a stronger, more theoretically-grounded link for my teaching-research nexus, and allowed some time and space for reflection on what—and how—I had been teaching.

13.8 Conclusions

Tourism higher education has a continuing role to play in ensuring a sustainable future; our students are important players in this. As Belhassen and Caton (2011, p. 1395) argue, 'the inclusion of [critical management studies] in university curricula carries intellectual, pedagogical, ethical, and professional added values that traditional approaches lack'. This chapter has outlined how CRP can provide a sound pedagogical and philosophical basis for teaching sustainable tourism. While the focus has been on education for sustainable development, other interested teachers in the tourism, hospitality and leisure fields could adopt similar strategies.

However, CRP is not an approach that can be instantly or easily subsumed; it requires time, effort and deep reflection, on the part of both the student and lecturer (Fullagar & Wilson, 2012). The CRP approach can also be confronting for teacher and learner, as our beliefs and values are made bare, our teaching practices are critiqued, our long-held assumptions are challenged, and the power relationship between teacher and learner becomes less hierarchical (Brookfield, 1987, 1995).

For these reasons, it is important to inform learners—and our own peers—of our critical intentions, and to allow for constant feedback and discussion.

It will be important to monitor how students respond to this CRP model over time, and to follow if and how their learning has been improved and impacted upon. This will particularly be the case for students from minority and/or international backgrounds, where requests for 'critical reflection' may not match cultural values and perhaps silence those groups (Brookfield, 1995). These will be important issues to consider in the future as my critical reflective practice continues.

Acknowledgement This chapter is based on some aspects of an earlier conference paper (Wilson, 2010), though it has been significantly changed and updated. The paper is Wilson, E. (2010, February 8–11). Practice what you teach: Using critically reflective practice in teaching sustainable tourism planning. In S. Crispin et al. (Eds.), *Proceedings of the 20th annual conference, Council for Australian University Tourism and Hospitality Education, Hobart, Tasmania*. Hobart: CAUTHE (CD Rom).

References

Atkinson, G., Dietz, S., & Neumayer, E. (Eds.). (2007). *Handbook of sustainable development*. Cheltenham: Edward Elgar.

Belhassen, Y., & Caton, C. (2011). On the need for critical pedagogy in tourism education. *Tourism Management, 3*, 1389–1396.

Biggs, J., & Tang, C. (2007). *Teaching for quality learning at university* (3rd ed.). Berkshire: McGraw-Hill/The Society for Research into Higher Education.

Brookfield, S. (1987). *Developing critical thinkers: Challenging adults to explore alternative ways of thinking and acting*. San Francisco, CA: Jossey-Bass.

Brookfield, S. (1995). *Becoming a critically reflective teacher*. San Francisco, CA: Jossey-Bass.

Brookfield, S. (2005). *The power of critical theory for adult learning and teaching*. Berkshire: Open University Press.

Brookfield, S. (2006). Developing critical thinkers: Supplementary materials. http://www.stephenbrookfield.com. Accessed 19 Apr 2007.

Cockburn-Wootten, C., Henderson, A., & Rix, C. (2005). Learning from the apprentice: An account of action research practice in a university department. *Action Learning: Research and Practice, 2*(1), 73–80.

Cuncliffe, A. (2004). On becoming a critically reflexive practitioner. *Journal of Management Education, 28*(4), 407–426.

Deale, C., & Barber, N. (2012). How important is sustainability education to hospitality programs? *Journal of Teaching in Travel and Tourism, 12*(2), 165–187.

Fisher, K. (2003). Demystifying critical reflection: Defining criteria for assessment. *Higher Education Research and Development, 22*(3), 313–325.

Fisher, K. (2004). *Critical reflection: What is it and how do you do it?* Paper prepared for Southern Cross University, Lismore, NSW (pp. 1–9).

Freire, P. (1970). *Pedagogy of the oppressed*. New York: Continuum.

Fullagar, S., & Wilson, E. (2012). Critical pedagogies: A reflexive approach to knowledge creation in tourism and hospitality studies. *Journal of Hospitality and Tourism Management, 19*(July, e2), 6 p. Published online 17/7/2012, http://dx.doi.org/10.1017/jht.2012.3.

Fulop, L. (2002). Practising what you preach: Critical management studies and its teaching. *Organization, 9*(3), 428–436.

Gabennesch, H. (2003). Critical thinking: What is it good for? (In fact, what is it?). *Skeptical Inquirer, 30*(2), 36–41.

Holmes, P., Cockburn-Wootten, C., Motion, J., Zorn, T. E., & Roper, J. (2005). Critical reflexive practice in teaching management communication. *Business Communication Quarterly, 68*, 247–256.

Hooks, B. (2003). *Teaching community: A pedagogy of hope*. New York: Routledge.

Hunter, C. (2002). Aspects of the sustainable tourism debate from a natural resources perspective. In R. Harris, T. Griffin, & P. Williams (Eds.), *Sustainable tourism: A global perspective* (pp. 3–32). Oxford: Butterworth-Heinemann.

Jafari, J. (1990). Research and scholarship: The basis of tourism education. *Journal of Tourism Studies, 1*(1), 33–41.

Jamal, T. (2004). Virtue ethics and sustainable tourism pedagogy: Phronesis, principles and practice. *Journal of Sustainable Tourism, 12*(6), 530–545.

Kearins, K., & Springett, D. (2003). Educating for sustainability: Developing critical skills. *Journal of Management Education, 27*(2), 188–204.

Merriam, S. B., Caffarella, R. S., & Baumgartner, L. M. (2006). *Learning in adulthood: A comprehensive guide* (3rd ed.). San Francisco, CA: Jossey-Bass.

Salazar, M. (2013). A humanizing pedagogy: Reinventing the principles and practice of education as a journey toward liberation. *Review of Research in Education, 37*, 121–148.

Schön, D. (1983). *The reflective practitioner: How professionals think in action*. New York: Basic Books.

Sterling, S. (2010). Living "in" the earth: Towards an education for our times. *Journal of Education for Sustainable Development, 4*(2), 213–218.

Wals, A. E. J., & Jickling, B. (2002). 'Sustainability' in higher education: From doublethink and newspeak to critical thinking and meaningful learning. *International Journal of Sustainability in Higher Education, 3*(3), 221–232.

Weaver, D. B. (2005). *Sustainable tourism: Theory and practice*. Oxford: Elsevier.

Welsh, A. M., & Murray, D. L. (2003). The ecollaborative: Teaching sustainability through critical pedagogy. *Journal of Management Education, 27*(2), 220–235.

Wilson, E. (2010). Practice what you teach: Using critically reflective practice in teaching sustainable tourism planning. In S. Crispin, A. Dunn, S. Fishwick, A. Franklin, D. Hanson, D. Reiser, R. Shipway, M. Wells, & C. Baxter (Eds.), *Proceedings of the 20th Annual Conference, Council for Australian University Tourism and Hospitality Education* (CAUTHE, Hobart, Tasmania, 8–11th February). Hobart: CAUTHE (CD Rom).

Wilson, E., & von der Heidt, T. (2013). 'Business as usual'? Barriers to education for sustainability in the tourism curriculum. *Journal of Teaching in Travel and Tourism, 13*(2), 130–147.

Chapter 14
Digital Immersion for Sustainable Tourism Education: A Roadmap to Virtual Fieldtrips

Christian Schott

Abstract Fieldtrips have been an important component of a range of educational disciplines for many decades and the associated pedagogies of active and experiential learning have been promoted since the early 1900s. Active learning, which is an integral part of fieldtrips has been found to act as a valuable means of engaging students with the subject, enhancing student's subject knowledge and understanding, and developing lifelong learners. Fieldtrips additionally allow for a concept or topic to be examined in its social, cultural, environmental and political context, thus creating a space for situated learning, which is recognised to assist in crystallising learning outcomes for students. The merit of fieldtrips to learn about complex topics, such as sustainable tourism development is thus well established. However, today's teaching and learning budgets are constrained and students are less able to pay for fieldtrips in the face of steadily increasing tuition costs in most countries than they may have been in the past. As a consequence, the tradition of fieldtrips has become less common despite its recognised educational value. In response, this book chapter presents a learning tool which provides a trade-off between the benefits and drawbacks of the comparatively affordable 'one dimensional' text based case study and the rich, authentic, but increasingly less accessible multi-dimensional experience of a real fieldtrip—by presenting a digital immersion (virtual) fieldtrip.

Keywords Virtual fieldtrips • Sustainable tourism development • Digital immersion • Active learning • Open Simulator • Fiji

14.1 Introduction

Fieldtrips have been an important component of a range of educational disciplines for many decades and the associated pedagogies of active and experiential learning have been promoted since the early 1900s by the educational philosopher John Dewey (1963). The active learning process which is an integral part of fieldtrips has been found to act as a valuable means of engaging students with the subject

C. Schott (✉)
Victoria University of Wellington, Wellington, New Zealand
e-mail: christian.schott@vuw.ac.nz

G. Moscardo, P. Benckendorff (eds.), *Education for Sustainability in Tourism*, CSR, Sustainability, Ethics & Governance, DOI 10.1007/978-3-662-47470-9_14

(Hanson & Moser, 2003; Schott & Sutherland, 2008), enhancing student's subject knowledge and understanding (Chickering & Gamson, 1987) and developing lifelong learners (Clegg, 2000, cited in Wolfe, 2006). While there is a lack of consensus over an exact definition (Prince, 2004), at its core active learning entails strong student ownership of the learning process and thinking about the things they are doing as part of that process (Bonwell & Eison, 1991), which differentiates it from passively focused "absorption learning". Fieldtrips additionally allow for a concept or topic to be examined in its social, cultural, environmental and political context, thus creating a space for situated learning, which is recognised to assist in crystallising learning outcomes for students (Jakubowski, 2003; Scarce, 1997). The merit of fieldtrips to learn about complex topics, such as sustainable tourism, particularly when situated in environments that are environmentally, socially, and culturally unfamiliar, is thus well established.

However, today's teaching and learning budgets are constrained, there are increasing workload pressures on staff (Dredge & Schott, 2013), and institutions are concerned about liability issues related to off-campus activities (Pearson & Beckham, 2005). Additionally, students are less able to pay for fieldtrips than they were in the past due to steadily increasing tuition fees, and there is growing recognition of the negative environmental impacts that are generated by air travel which can be part of fieldtrips (Schott, 2012). As a consequence, the tradition of fieldtrips has become less common despite its recognised educational value. The growing cohort of educators who don't have the option of incorporating a fieldtrip into their curriculum, but nevertheless wish to deepen students' learning by applying complex concepts to a meaningful context, are faced with two very different alternatives:

1. Revert to predominantly text-based case-studies which describe the case's context and complexities with the support of illustrations and in some cases videos; or
2. Embrace twenty-first century technology and harness its educational capabilities by developing a tool for 'digital immersion' active learning using virtual reality software.

While text-based case studies have in recent years become less one dimensional and descriptive through the support of web-based illustrations, the richness of both active and situated learning is largely absent in this learning tool. Equally, the ability for students to learn about the profound interrelationships that are central to complex topics, such as sustainable development, is compromised by text-based case-studies. The second option, which can best be described as a virtual fieldtrip, has the ability to 'bring alive' the information contained in a text-based case study in addition to replicating some of the complexities and contextual richness of real fieldtrips by providing a platform for (digitally immersed) situated and experiential learning. The need for greater use of non-text based learning tools, is further underscored by the diversity of learning styles present in our classrooms; our current generation of students is more diverse than ever, which inevitably leads to an increased prominence of a diverse range of learning styles (such as kinaesthetic, spatial-visual and audio), in addition to text-based learning styles. As such, the virtual fieldtrip can be conceptualised as a trade-off between the benefits and drawbacks of the comparatively affordable 'one dimensional' text based case study and the rich, authentic, but increasingly less accessible multi-dimensional experience of a real fieldtrip. Based on this strong pedagogic rationale and given that many students have experienced a digitalised childhood and youth, this chapter promotes the digitally immersed case study (referred to as the virtual fieldtrip), as an effective learning tool that aligns strongly with the digitalised everyday lives of students.

14.1.1 Conceptualising 'Virtual Fieldtrips'

Over the last 15 years educators spanning many disciplines have been inspired to recreate aspects of the 'field' in visual and audio format with the aim of adding meaning and context to the topic being discussed. Some refer to these learning tools as 'virtual fieldtrips'; however, the term is used for a wide variety of teaching and learning tools. Examples include videos of a teacher's real fieldtrip, pictures of a particular location supported by interpretative text, use of Google Earth coupled with YouTube videos and 360° images to provide a visualisation of a place and its features, and the most recent projects which present virtual reality 3D renderings of buildings or specific reality-inspired objects. All of these approaches to enrich the teaching and learning experience are valuable as they appeal to a variety of learning styles and seek to present an authentic account of a case. However, current technology allows a much greater penetration into the foundations of experiential and situated learning by immersing learners in a complex, three-dimensional learning space. What sets this project apart from the various 'virtual fieldtrip' formats outlined above are: the experimental recreation of crucial 'human and cultural facets' of the real world; the large scale of the case which incorporates two villages; the nature of the assessments (fieldwork) requiring groups to learn by conducting research in the field, and finally the ability of the students' group discussions to take place 'in the field' (between avatars while in the virtual environment). For this project, a virtual fieldtrip was thus conceptualised in the following way:

> A virtual fieldtrip recreates a wide range of aspects and complexities of the real world in digital format using both audio and visuals (3D objects, videos, still images, and documents). It allows student researchers to digitally immerse themselves and collaborate in the environment (physical and social/cultural) with the aim of completing fieldwork tasks similar in nature to those set during real fieldtrips.

The project was designed for second year Tourism Management students at Victoria University of Wellington (New Zealand) with funding from the VUW Teaching and Learning Development Fund. It was specifically targeted at a compulsory second year course on Sustainable Tourism Development (STD) because of the myriad of complexities that are inherent to STD but difficult for students to truly appreciate while sitting in a lecture theatre. The virtual fieldtrip seeks to deepen students' learning by applying the theory from lectures and readings to two communities on a remote Fijian island, which are faced by both local and global challenges that most students in New Zealand struggle to relate to; but which are not uncommon in developing countries. Additionally, Fiji is important from a tourism management perspective as it is the third most visited country by New Zealanders (Ministry of Business, Innovation and Employment, 2013) and as such a place of great relevance to New Zealand undergraduate Tourism Management students—as future managers as well as tourists. While providing an authentic case for students to apply the concept of STD to the realities of a community, learning about these realities also serves to reinforce the acute need for STD which in turn illustrates the relevance of the topic and the rationale for learning about it. Three specific pedagogical drivers underpin the project:

- A desire to enhance authenticity of the learning context;
- Fostering an acute sense of the interconnectedness and complexities of social, cultural, environmental and economic factors impacting communities (particularly in developing countries); and
- Increasing access to, and passion for, learning

14.1.2 The Fieldwork Task

The student task was designed as a role play where the second year students worked for a New Zealand-based Tourism Consultancy which specialises in STD. Students worked in project groups of three or four and were dispatched to conduct fieldwork on the (virtually recreated) Fijian island in response to this fictional initiative:

> A collaborative initiative between the Government of Fiji and NZ Aid has been set up to determine whether strategic STD of a remote Fijian island would result in the positive impacts outweighing the negative ones—the three dimensions of sustainable development need to be considered as do the direct, indirect and induced, and the short as well as long term impacts.

As students needed to determine whether the benefits of STD would outweigh the associated costs, they had to incorporate compromises and trade-offs in their decision-making; much like in the 'real world'. Based on this process, groups could either (a) devise a STD plan for an area of their choice, or (b) present an argument against any development initiatives. To combine deep learning with the development of diverse communication skills, a presentation of the proposal was followed by the submission of a detailed group report in the form a wiki (collaborative, media enhanced online document).

To render the fieldwork experience as authentic as possible students were advised to start their visit at the same point where the researchers who developed the virtual fieldtrip arrived on the real island. Fijian cultural protocol dictates that on arrival visitors must meet the village elders and present gifts (kava) as part of the

'I sevusevu' ceremony, which serves to explain the visitors' intentions and to seek welcome onto the vanua's (community) land. Students need to watch the video of the researcher's 'I sevusevu' ceremony to understand this important Fijian custom, which is the first of a number of cultural insights which emphasise both the significance and nature of the vanua's cultural protocols. Afterwards, the students freely explore to learn about the island: the geography, topography, flora and fauna, the two villages and their layout, the communities' customs, community members' opinions and aspirations, any potential touristic resources, the current sourcing of fresh water, how the communities meet their energy needs, schooling, and health facilities amongst other important considerations.

Based on this fieldwork each project team had to initially discuss and decide whether they felt that the positive impacts of any STD would likely outweigh the associated negative impacts. There are no 'right' or 'wrong' answers to this, rather the task is about presenting an informed argument. The teams that worked on a proposal incorporating a sustainable tourism development plan had to address five key criteria: products & facilities, target market, community participation/consultation, maximisation of benefits for community, and natural resources & the environment. The teams that were opposed to any development were asked to discuss the rationale behind their decision by addressing the economic, environmental, and socio-cultural reasons. Following a group presentation of the pro or contra proposal to the fictional funding bodies a more detailed proposal was submitted by each group using a wiki, thus allowing for easy collaboration and extensive media enhancement of the proposals.

14.2 Developing a Virtual Fieldtrip: A Development Roadmap Based on the Fijian Virtual Fieldtrip Project

To illustrate the process of developing the virtual fieldtrip, as briefly described above, key steps will now be outlined for those who are interested in replicating the concept, which can be applied to a variety of settings and topics; but is a particularly powerful learning tool for sustainable tourism because of its complex and applied nature. The selection process for suitable technology and software was conducted in mid 2011. As such, the software selection process for a replication project should be conducted using the options available at that time rather than relying on the options identified for this project in 2011.

14.2.1 Software Selection Process

All potentially suitable software was identified and evaluated against a set of criteria including: user friendliness, adaptability, versatility, interactivity, and a number of other specific points. For each of the criteria it was considered important to distinguish between three different project dimensions: the initial building phase, the ongoing maintenance, and the daily use by learners (see Appendix for more detail about the criteria and the scoring matrix). This resulted in a short list of three possible programs that were further evaluated.

The software program ultimately selected for the project was the Open Simulator (OpenSim) because of the following key benefits: in step with progress in virtual world software and appeal to users, no development or usage charges, no server/ infrastructure costs because of ability to join a New Zealand-based OpenSim digital learning initiative called the New Zealand Virtual World Grid (primarily used for role play and architectural design), software and hardware requirements were met by VUW student computers, ease of use and speed of development functions, and the ability to update the virtual world.

In addition to OpenSim software which runs the virtual world, a 'viewer' is required. While several viewers are available we found that Imprudence Viewer served the project needs best. It is also worth noting that OpenSim requires Internet access to constantly download the virtual world's content; the related data usage varies between roughly 40 MB of data per hour for low resolution settings, and roughly 200 MB per hour for the highest resolution settings.

14.2.2 Scale of the Virtual Island

The creation of the project began with a detailed assessment of the Fijian island's size relative to server capacity and practicality considerations in order to determine how many 'regions' would be appropriate to develop the virtual island. At the same time, consideration was given to the scale at which the island and its detail would be created as the relative size of objects and distances between villages and facilities are important features of a virtual fieldtrip. After discussion with the server manager, the decision was taken to develop the virtual island at a scale of roughly 1:3. At this scale the virtual island was spread over eight regions with the most notable exception to the 1:3 scale being applicable to the two regions connecting the two villages where only an approximately 1:6 scale could be achieved (see Fig. 14.1); while the change in scale is not accurately reflective of the island, the key messages about significant distances (such as the time required for children to walk to school) were still clear through the use of signs. The benefits of developing the virtual island at this scale and in this manner are: (1) that while all the significant detail and information was still placed in its appropriate context it was believed that students are less likely to lose interest in exploring the island than if it took three times as long to walk from point A to point B, and (2) that lesser demands are placed on the server than if more regions were utilised for the virtual island.

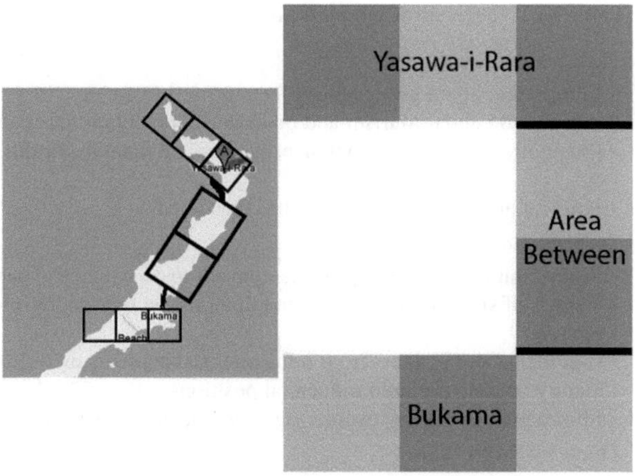

Fig. 14.1 Map of Bukama/Yasawa-i-rara and general layout of the server regions

14.2.3 In-Situ Fieldwork to Collect Required Information

The author, and a cultural and linguistic adviser from VUW's Professional and Executive Development Program, Maciu Raivoka, travelled to the Fijian island in June 2011 to gather all the required information for the project. In order to develop the virtual island as an effective learning tool the project enlisted a framework commonly used to develop text-based case studies. The framework is adapted from Tyson (2011) who is the principal case writer for the Australia and New Zealand School of Government Case Program, which holds in excess of 160 case studies. The framework (Table 14.1) was adapted to the needs of the project by focusing on (a) a digital-immersion presentation of the case (rather than a structured textual account), (b) aspects relevant to a decision-forcing case (rather than a concept-application or illustrative one), and (c) the specific context of a sustainable tourism case. It is worth noting here that due to word count restrictions this roadmap is only able to document the project's key decision-making steps alongside some technical information; the important cultural dimension can unfortunately not be covered in this short case-study chapter.

A range of media were employed to collect and document the information set out in the case planning framework; some was used in the virtual world itself, while other information informed the digital design of the environment and objects. The range of media included:

- Videos with audio
- Digital Photographs
- Digital and hard copy documents (e.g. tourism statistics, government reports)
- Digital map files

Table 14.1 The case planning framework

Background—macro
• Information about geographical location and features • Information about tourism and other development on the island • Relativities: e.g. populations size, economic dynamics, health care, education, role of religion, etc. • Political and cultural aspects to be considered
Background—more specific
• Recent initiatives and/or projects relevant to Sustainable Tourism Management • Variety of stakeholders' views and opinions on tourism development and development more broadly • Comments and insights from influential stakeholders as well as from members of the community who do not hold influential positions • Information about key events: e.g. new legislation, natural disasters
Dilemma or decision
• Any dilemmas need to be identified and related material needs to be collected • Develop the range of key decision avenues: examining the factual aspects from a disciplinary perspective and by listening to the voices of a wide range of stakeholders • Identify and collect information which is authentic and provides support for the identified decisions avenues

Adapted from Tyson (2011)

Fig. 14.2 Example of early stage of creating the island

14.2.4 Designing the Virtual Island

Using a map of the real island as a guide, each of the eight regions was shaped to reflect the basic geography and topography of the real island; see example in Fig. 14.2.

Subsequently buildings and vegetation were added. Most of the buildings had to be specifically created for this project and could not be copied from a digital object library as the distinctive identity of the island had to be retained. Such buildings included the chiefly houses, the community hall, the Church Minister's house, island bures, and school buildings amongst others (see Fig. 14.3).

Fig. 14.3 The Yasawa-i-rara community hall, both real and virtual

14.2.5 *Infusing the Virtual Island with Real-World Videos and Photographs*

Videos were placed in the same spot on the virtual island where they were initially recorded. They were displayed as a framed image of the person talking in the video (Fig. 14.4). When users left clicked on the framed image, the video appeared in an in-world web browser. The video files were stored on 'YouTube' which allows for free storage and streaming of the videos. However, because the people in the videos requested for them not to be publicly available (for educational purposes only), the video files were categorised as 'unlisted', which restricts access to only those who have the URL. To also represent the valuable tourists' perspectives of the island, two YouTube videos posted by tourists illustrating how they experienced the island were also incorporated. These videos portrayed the typical tourist perspective (arriving by small plane and by cruise ship), which provided a valuable contrast to the community perspectives. Additionally, a tourist who was staying on the island was video-interviewed about his views relating to both the island itself and tourism on the island. The intention was to video-interview a larger number of tourists, however, significantly less tourists than expected were on this remote island at the time of the visit.

A few photographs were also used integrated into the virtual island to either emphasise the similarity between the virtual world and the real world, or to illustrate a particular scene/activity which was very challenging to meaningfully replicate in the virtual world (such as a woman washing clothes in a bucket to illustrate the scarcity of water and electricity, or tropical fish on a coral reef to illustrate the pristine and abundant marine life surrounding the island). Photographs

Fig. 14.4 Example of integration of videos in-world

were also positioned in the same area where they were taken and stored digitally on an image hosting server (ImageShack.us), which provides this service at no cost.

14.2.6 In-World Communication Between Users

OpenSim offers an inbuilt text chat system in which users can send messages to each other through the use of a text field. This function was considered important to allow groups of students to collaborate on the fieldwork tasks while in-world. An additional benefit was that they can collaborate in world in this manner even when group members are at their home computers several kilometres apart. Users could create groups, allowing them to have a discussion with only their group members. The text chat function also proved very useful during the island's design phase as it allowed the project leader to communicate specific instructions about placement of objects to the technical RA in-world.

14.3 Problems and Challenges Encountered

After the project had been piloted as part of the Sustainable Tourism course the virtual island was attacked by a hacker. Significant damage was inflicted as numerous random objects were placed on the virtual island and bugs inhibited normal movement of the avatars. As a result, the program could not be used for a period of 6 weeks while the vandalism was rectified. The initial solution was to clean the virtual world of any foreign objects and bugs. However, this cleaning process was unsuccessful in two regions which lead to the decision to replace the two regions with earlier archived copies. One approach for countering such attacks is to increase security by restricting access to the virtual island; although it is unclear whether this cyber vandalism attack could have been prevented by this security measure.

As can be expected from a pilot project of this nature a number of more specific problems were also encountered; suggested solutions are also listed:

- When first using the program students were removing objects in-world (mostly unintentionally) as well as creating 'random' objects. This occurred when students right clicked (for example to activate a video) instead of left clicking.
  Solution: ensure 'build' privileges are removed from all student users, anchor any objects, and educate students about the need to be careful about any changes and the impacts they can cause.
- Instabilities in the program's or specific region's accessibility were noted on several occasions.
  Solution: the problem either self-resolved when it was network related, or more commonly, was quickly resolved by the technical team hosting the server.

14.4 Evaluation and Conclusions

After using and refining this digital immersion learning tool for 2 years the project team strongly believes that virtual fieldtrips, both as a concept and as implemented in this project (using OpenSim), present a very valuable and effective learning tool. This assessment is based on: (1) the project team's own pedagogical assessment (2) the nature and level of student engagement with this learning tool observed during the workshops, (3) the quality of the virtual fieldwork and subsequent proposals presented by student groups, and (4) informal student feedback about the virtual island as a learning tool.

However, this overall positive assessment of the project should not be understood to suggest that either the program or the related student fieldwork tasks are 'flawless'. The process of reflection and further refinement will continue for many years and, where time and financial support permit, further improvements to the virtual island will be made. As such, a more recently developed software platform from the game design domain is currently being explored for this project due to its higher quality visuals.

Based on the experience documented, we recommend that greater use of digital immersion should be encouraged for learning in general, and for learning about complex and applied concepts such as sustainable tourism development in particular. However, to embrace and harness these new and exciting digital opportunities for higher education, institutions need to actively support the development of digital learning tools. In addition to student computers meeting software requirements, three broader levels of support are needed from the institution: provision of hardware if Internet-based (servers to host virtual island), skilled development of the virtual world which can be multi-functional to serve the needs of different courses (shaping of geography, topography, building of objects, etc. according to specific requirements), and ongoing availability of technical support staff with digital design and network skills. Although a collaborative approach to sharing hardware and technical skills across institutions has a number of strengths, the provision of a server and availability of technical skills within each institution using virtual fieldtrips can assist greatly in minimising the amount and severity of problems that may occur. Although the funding needs required to design a digital immersion learning tool for a single course may appear high, there is significant potential to lower the per course costs. This can be achieved by allowing a variety of courses to use a common virtual learning environment and for digital immersion learning to be used across a variety of disciplines, which would rapidly justify the hardware and technical skill investment. With increasing recognition that the educational experience of current higher education students needs to align more solidly with twenty-first century technology, it is only a question of time before such digitally immersed, situated and active learning experiences are embraced by a wider range of tertiary education institutions; in particular for critically important, yet complex topics such as sustainable development. Technology will continue to innovate rapidly, leading to more refined digital immersion experiences over the coming years and decades. However, at the time of writing OpenSim software presents an effective introductory option to allow students to learn about the sobering realities of life in developing countries, the positive and negative impacts of tourism, and the inherent interconnectedness of environmental, socio-cultural and economic factors—in other words as an effective tool for sustainable tourism education.

Video Illustration of Virtual Island To view an 11 min video illustrating basic features of the learning tool and selected regions of the virtual island please click on this URL: http://youtu.be/yjSIONrRhmo. But it needs to be noted that the technology is of a 2012/13 standard, and that the richness and comprehensive nature of this learning tool as well as a fuller sense of digital immersion is best experienced when controlling one's own avatar on the virtual island.

Acknowledgements I wish to thank the VUW Teaching and Learning Development Fund for supporting this pilot project. Additionally, I am grateful to the people of Yasawa-i-rara and Bukama for their wonderful hospitality and for kindly sharing their views and experiences of life on a remote South Pacific Island. As project leader I would also like to thank the other project members for their valuable expert contributions: Maciu Raivoka (VUW's Professional and Executive Development Program); Prof Warwick Murray (VUW Development Studies); David McLean (Freelance Digital Designer); and Dr Scott Diener and his Digital Design Team at Auckland University.

Appendix

Assessment criteria Scores from 1 (lowest) to 5 (highest)	Building (developer)		Maintenance (instructor)		Daily use (students)	
	Score	Comments	Score	Comments	Score	Comments
User friendly (overall)						
• Accessible						
• Attractive						
Adaptable (overall)						
• Accommodates various media formats						
• Capacity to develop large amount of material and detail						
• Expandability (students able to add content in separate area)						
• Scale of space (can island 10 km × 1 km be replicated?)						
Versatile						
• Useable by different disciplines						
• Cost (start up and annual fees)						
• Time effectiveness						

(continued)

Assessment criteria Scores from 1 (lowest) to 5 (highest)	Building (developer)		Maintenance (instructor)		Daily use (students)	
	Score	Comments	Score	Comments	Score	Comments
Interactivity (overall)						
• Can learners communicate while 'in world'?						
• Can interactive datasets (modifiable by user) be incorporated?						
• Level of user control (teleportability, movement control, etc.)						
Other (delete as appropriate)						
• Longevity & accessibility of software (around in 5 years?)	100 %	75 %	50 %	25 %	0 %	
• Can access be restricted?		Yes	No			
Computer specifics						
Demand on computer specs (graphics card/processor)	Comments:					
Demand on bandwidth (if 80 students using at same time)	Comments:					
Demand on bandwidth (if 40 students using at same time)	Comments:					
Other (added during development process)						
Usage of internet data (important for use of home computers)	Comments:					
Ease of installation on home computers	Comments:					

References

Bonwell, C., & Eison, J. (1991). *Active learning: Creating excitement in the classroom.* San Francisco: Jossey-Bass.

Chickering, A., & Gamson, Z. (1987). Seven principles for good practice in undergraduate education. Retrieved May 6, 2008, from http://www.uis.edu/liberalstudies/students/documents/sevenprinciples.pdf

Dewey, J. (1963). *Experience and education.* New York: Collier (originally published 1938).

Dredge, D., & Schott, C. (2013). Academic agency and leadership in tourism higher education. *Journal of Teaching in Travel and Tourism, 13*(2), 105–129.

Hanson, S., & Moser, S. (2003). Reflections on a discipline-wide project: Developing active learning modules on the human dimensions of global change. *Journal of Geography in Higher Education, 27*(1), 17–38.

Jakubowski, L. M. (2003). Beyond book learning: Cultivating the pedagogy of experience through field trips. *The Journal of Experiential Education, 26*(1), 24–33.

Ministry of Business, Innovation and Employment. (2013). Tourism research and data. Retrieved July 26, 2013, from http://www.med.govt.nz/sectors-industries/tourism/tourism-research-data
Pearson, D., & Beckham, J. (2005). Negligent liability issues involving colleges and students: Balancing the risks and benefits of expanded programs and heightened supervision. *Journal of Student Affairs Research and Practice, 42*(4), 796–1173.
Prince, M. (2004). Does active learning work? A review of the research. *Journal of Engineering Education, 93*(3), 223–231.
Scarce, R. (1997). Field trips as short-term experiential education. *Teaching Sociology, 25*(3), 219–226.
Schott, C. (2012). Virtual mobilities and sustainable tourism: Virtual fieldtrips for climate change Education. In T. R. Tiller (Ed.), *Conference proceedings of BEST EN think tank XII* (pp. 341–356). Sydney: University of Technology Sydney.
Schott, C., & Sutherland, K. (2008). Engaging tourism students through multi-media teaching and active learning. *Journal of Teaching in Travel & Tourism, 8*(4), 351–371.
Tyson, J. (2011). Personal communication. Wellington, June 2011.
Wolfe, K. (2006). Active learning. *Journal of Teaching in Travel & Tourism, 6*(1), 77–82.

Ministry of Business, Innovation and Employment (2013). Tourism research and data. Retrieved July 28, 2013, from http://www.med.govt.nz/sectors-industries/tourism/tourism-research-data

Chapter 15
Global Programs in Sustainability: A Case Study of Techniques, Tools and Teaching Strategies for Sustainability Education in Tourism

Michael A. Tarrant, Lee Stoner, Kristin Tessman, Mikell Gleason, Kevin Lyons, and Stephen Wearing

Abstract This chapter describes the pedagogical approach adopted by the Discover Abroad's Global Programs in Sustainability (GPS) and its growth from a concept in 1999 to the largest single education abroad program for students at UGA and one of the most popular in the country. Focusing on broad questions of human–environment sustainability, GPS uses a module-based pedagogical approach to deliver inter-disciplinary, faculty-led, study abroad programs for students from virtually every major on campus to a range of destinations. The mission of GPS is to foster humility through programs that address sustainability, conceptualized as the "meaning of progress". Quantitative evidence of GPS student learning outcomes has been published in several leading journals and suggests that a one-size-fits all approach cannot be justified in study abroad. Rather than encouraging students to simply go abroad, academic advisers should attend to the needs of students beyond their desired country of destination and academic goals, to include professional development including higher-order outcomes such as global citizenry.

Keywords Study abroad • Global programs • Sustainability • Educational travel

---

M.A. Tarrant (✉) • K. Tessman • M. Gleason
University of Georgia, Athens, GA, USA
e-mail: Tarrant@uga.edu; kristinr@uga.edu; mikell.gleason@gmail.com

L. Stoner
Massey University, Wellington, New Zealand
e-mail: dr.l.stoner@gmail.com

K. Lyons
University of Newcastle, Callaghan, NSW, Australia
e-mail: Kevin.Lyons@newcastle.edu.au

S. Wearing
University of Technology Sydney, Ultimo, NSW, Australia
e-mail: Stephen.Wearing@uts.edu.au

G. Moscardo, P. Benckendorff (eds.), *Education for Sustainability in Tourism*, CSR, Sustainability, Ethics & Governance, DOI 10.1007/978-3-662-47470-9_15

15.1 Introduction

Educational travel, including study abroad, has become a major form of travel and tourism with 274,000 U.S. students studying abroad annually (an increase of over 300 % since 1994) and millions of students worldwide traveling overseas for education (Institute of International Education, 2012). This chapter describes the pedagogical approach adopted by the Discover Abroad's Global Programs in Sustainability (GPS) housed in the Warnell School of Forestry and Natural Resources at the University of Georgia (UGA) in the United States, and its growth from a concept in 1999 to the largest education abroad program at UGA (and one of the largest in the U.S.). Each year since 2006 GPS has worked with between 250 and 350 students, contributing to UGA ranking in the top five of all Research I universities nationally for the number of students on short-term study abroad programs (Institute of International Education, 2012). While student numbers have been important (with more than 2500 students since inception and representing almost 15 % of all UGA students who study abroad each year), equally important has been the breadth of faculty involvement (from assistant professors to college deans) and academic college/departmental support (with 12 of the University's 16 colleges and schools collaborating with GPS).

Focusing on broad questions of human–environment sustainability, GPS uses a module-based pedagogical approach to deliver inter-disciplinary, faculty-led, study abroad programs for students from virtually every major on campus to a range of destinations (see www.DiscoverAbroad.uga.edu). The GPS approach, initially adopted by consortia representing over 20 U.S. colleges and universities in 2006, has resulted in an additional 500 students or more annually. For example, from 2008 to 2010, approximately 4.7 % of all U.S. students studying abroad in Australia and New Zealand participated in one of the UGA or consortium programs. Since 2008, an ongoing large-scale empirical research effort has addressed student learning outcomes, resulting in several publications with leading travel/tourism and international education journals (e.g., Tarrant, 2010; Tarrant & Lyons, 2012; Tarrant et al. 2014; Tarrant, Rubin, & and Stoner 2014; Tarrant et al., 2011; Wynveen, Kyle, & Tarrant, 2012).

15.2 Context of the Activity

The concept of GPS began in 1999 when representatives of five UGA academic units gathered to create a single course in sustainable development. At the request of a former UGA Vice-President for Instruction, our task was to propose an approach for combining four UGA core required courses: Introduction to Anthropology, Environmental Issues, Introduction to Human Geography, and Introduction to Global Affairs. The result was (1) a single study abroad course in Sustaining Human Societies and the Natural Environment, cross-listed in six academic units (anthropology, ecology, education, forestry and natural resources, geography, and international affairs) at both the undergraduate and graduate levels and (2) a one semester, study abroad program that molded the four core required courses using a module-based curriculum approach.

The mission of GPS is to foster humility (i.e., a love and respect for the diverse peoples of the world and the environment in which we live and belong) through programs that address sustainability (conceptualized as the "meaning of progress"). We focus on questions of how humans interact with the natural environment, with a view to understanding how we may live more sustainably, and extend this guiding philosophy to broader disciplines. In understanding progress, students are required to consider the range of values (economic and non-economic) that are critical to expanding their moral horizons. We believe that sustainability cannot be considered outside of social equity (or a Green Paradigm) that acknowledges a movement away from a Dominant Western Paradigm (focused on economic growth as the sole/core indicator of progress). Consequently, students are challenged to question whether there are other ways of doing things, of living sustainably, and maintaining a healthy and progressive standard of living through experiencing other cultures and countries.

As an example of this paradigm shift, social equity has arguably been a political agenda in New Zealand for many years, dating back to women's suffrage (New Zealand was the first country in the world to give women the right to vote in 1893) to the recent anti-smacking legislation in 2009 (protecting the rights of children) and, moving forward, in a proposal to become the world's first smoke-free country by 2025. Unlike many other European colonized societies, the first peoples (the Maori) have been represented in many areas of society resulting, in part, from the Treaty of Waitangi in 1840 and, more recently, from subsequent court settlements through the Waitangi Tribunal (e.g., Office of Treaty Settlements, 2008). In sum, study abroad in New Zealand provides U.S. students a place to study a form of progress that has moved beyond the Victorian ideals of land as a commodity to a Green philosophy in which human–environment relations span a range of non-economic and economic values and includes, at the core, the concept of *kaitiakitanga*, an obligation to take care of, and to protect, our natural resources as guardians of their future.

As global citizens, Americans are recognizing the need to balance economic, social, and environmental demands. Issues such as human population capacities, climate change, biodiversity preservation, and environmental pollution transcend national boundaries and our responses will accordingly need to be international and global in perspective. Furthermore, such problems and their solutions not only have complex ecological and biophysical bases but they are also dependent on understanding the social, cultural, historical, and political contexts. As a result, we adopt an integrated, multi-disciplinary approach that is relevant for students of most, if not all, majors. Our greatest strength has been in improving student access to programs by bridging diverse (often times, seemingly unrelated) academic units to (a) consider their disciplinary perspectives and relations to sustainability and (b) explore how their disciplines could work with other subjects to form an interdisciplinary study abroad program in sustainable development. For example, Education, English, History and Forestry have created a program on *British Landscapes and Literature*, while Theater, Anthropology, and Ecology teamed up for a program in India titled *Nature and the Human Spirit*.

A primary goal of GPS is to explicitly relate sustainable development issues across multiple disciplines. Subsequently, course credit for GPS is/or has been available in at least the following majors: Agriculture and Applied Economics, Anthropology, Art, Biology, Business, Comparative Literature, Ecology, English, Epidemiology, Food and Nutrition, Forestry and Natural Resources, Geography, Geology, Health Policy and Management, History, International Affairs, Journalism and Mass Communication, Literacy Education, Marine Sciences, Physical Education, Public Health, Recreation and Leisure Studies, Social Science Education, Speech Communication, and Theater. Formerly titled Studies Abroad in the South Pacific (and focused on the Australasian region exclusively), programs are now offered across the globe, in Antarctica, Australia (multiple states), Bali and Lombok, Fiji, India, Ireland, New Zealand, Tahiti, and the United Kingdom. Programs range from short- (1–4 weeks) to mid- (one semester) length and provide credit in university core required courses, major core courses, and electives. While most students are undergraduates, a growing number of students are graduates and/or take honors (graduate) level courses.

15.3 Description of the Teaching Activity

Two key aspects have characterized the GPS pedagogical approach: (1) adoption of a module approach and (2) an operational capacity to work with students and faculty from multiple disciplines. The GPS curriculum approach is built around faculty-devised modules, each of which relates to a specific theme and consists of (a) an introductory/background narrative, (b) direct instruction (field experiences/ observations, UGA/host faculty classroom lectures, informal conversations with faculty, small group seminars, and meetings with specialists and professionals), (c) related readings, and (d) assessment (comprised of ~250-word essays about complex ecological, environmental and social issues; peer-reviews; group debates; and digital stories). Not only is this a writing-intensive approach that meets the UGA Writing Certificate standard, but it incorporates strong social- and multimedia elements that recognize (and build upon) the real world in which students live and learn.

Our operational approach provides a comprehensive delivery mechanism for the academic, administrative, and logistical development and implementation of (primarily short-term) faculty-led programs on sustainable development, by facilitating collaboration across diverse (and sometimes seemingly disparate) academic units. (According to the 2012 Institute of International Education Open Doors Report, short-term programs represent the fastest growth sector currently attracting almost 60 % of all education abroad students.) As such, GPS provides an umbrella framework for (a) developing academic material; (b) faculty preparation (risk assessment, health and safety training, and program policies/procedures); (c) student recruitment; (d) student preparation (online pre-departure orientation, in-country orientation, and Program Handbook); (e) conducting all in-country logistics; and (f) a collaborative research study on student learning outcomes and global citizenry.

15.3.1 Module Approach

The module approach is unlike most other approaches to teaching and learning that students have experienced on campus. In typical campus classes, students learn through lectures in a somewhat linear fashion with one class building upon another. In contrast, ours is a holistic approach, akin to a mosaic, in which the complete picture only gradually comes into focus as more and more pieces of the mosaic are put into place. When the last piece of the mosaic is in place, the picture is complete, depicting the complex and multifaceted nature of what has been created. To push this analogy a little further, the pieces of the mosaic are like pieces of information, and the complete mosaic is the knowledge that has been gained of the subject. Students typically find the module approach confusing and even frustrating early on. Where does one find the pieces of information? Where does this piece fit? Does this piece fit? How does this piece relate to the overall topic? The single greatest advantage, however, is that it obliges students to be an active participant in the learning process and actively engaged in finding the pieces of information from multiple sources. In practice, this means listening and looking, taking good notes, asking good questions, and generally taking advantage of all of the resources and opportunities they encounter. It is a way of learning that is far removed from the taking and regurgitating of lecture notes.

This approach is novel and challenging for many students but, as evidenced from our recent studies (e.g., Tarrant & Lyons, 2012; Tarrant, Rubin, & Stoner 2014; Tarrant et al., 2011) most complete the program with a renewed worldview and a new lens for addressing critical socio-scientific issues. It is a writing-intensive approach to study that requires students to have a clear understanding of the question, to develop a central message/thesis that answers the question, and to formulate a response in a grammatically correct, concise, and non-redundant manner (in ~250 words only or in 6 min debates). We believe that reading, writing and communication skills are critically important to gaining the most from higher education and for being successful in the workplace. Those who speak and write well are at a clear advantage when competing for jobs and promotions.

The programs conclude with a requirement that students submit a 3-min digital story addressing the following question: How has your understanding of progress (and your personal values) changed, if at all, as a result of this program? The stories enable students to address the "big picture" of their study abroad experience relative to the academic goal of progress by projecting their own (substantiated) values and opinions in a social media format/delivery mechanism. The approach builds on the theory of transformational learning (Kegan, 2000; Mezirow, 2000) by asking students to reflect on their experience using the following cues/guiding questions (adapted from Dolby, 2007):

- What did you learn about the country that you visited?
- What did you learn about yourself, as an American?
- How have your perspectives about the world changed?
- What, if any, changes in your own behavior do you anticipate as a result of this program?

15.4 Evaluation of the Activity

In 2012 GPS received an Honorable Mention by the Institute of International Education (IIE) Heiskell Award for Innovation in International Education (http://www.iie.org/Who-We-Are/IIENetwork-Membership/Heiskell-Awards/University-of-Georgia). Considered to be the pre-eminent award for study abroad in the country, GPS is the first (and to date, the only) UGA program to be honored with such. For many students and faculty, the impact of their engagement with GPS has been transformational. Faculty collaboration has resulted in a substantial growth in the number and range of courses being offered across campus (and in the breadth of faculty involved); while, student evaluations acknowledge the growth in personal confidence as well as professional achievements:

> This trip has forced me to break down who I am and my beliefs which has allowed me to understand my prejudices and see the world from a new perspective with a new understanding of what is culture...I have grown so much as a person and my mind has been opened to a level of global thinking that I never knew existed... Experiences that I will use every day of my life for the rest of my life.

Equally as important as student comments are the evaluations of faculty. Dr. Alexandra Brewis-Slade (Executive Director and Professor of School of Human Evolution and Social Change at Arizona State University) acknowledges that:

> It is very challenging to develop programs that engage students of all backgrounds and inclinations, but this has been one of the reasons GPS has been so successful.... [the programs] are driven educationally by a vision of promoting lifelong environmental stewardship in UGA students as the future Georgia and US citizenry, and by all assessments achieve this elegantly and effectively.

State University of New York's Environmental Science and Forestry Provost Dr. Bruce Bongarten notes:

> My personal perception of what study abroad courses can be was completely altered. I saw students' eyes and minds opened. I saw curiosity and excitement for learning renewed. I saw students grow intellectually and personally within the course of a single month.

In evaluating the module approach, Dr. Steve Elliott-Gower (Director of the Honors Program, Georgia College) proposes, "it is an approach that maximizes the benefits of a study abroad experience."

Quantitative evidence of student learning outcomes (growth in global citizenship and pro-environmental behavior) has been published in several leading journals of international education and in travel and tourism. A description of the theoretical (i.e., Values-Beliefs-Norms) framework (providing a conceptual rationale for why study abroad can promote learning outcomes) can be found in Tarrant (2010). The empirical work has demonstrated that engagement in GPS fosters global citizenry by promoting critical assessment of justice issues (as compared with, for example, lower-level citizenry such as personal responsibility and participatory citizenship) with the effect of the program varying by (1) destination/country, (2) gender, and (3) previous study abroad experience (Perry et al., 2013; Perry, Stoner, & Tarrant, 2012; Tarrant & Lyons, 2012; Tarrant, Rubin, & Stoner 2014; Tarrant et al., 2011; Wynveen et al., 2012). Our current work suggests that studying abroad is associated with higher global perspectives (one of several desired learning outcomes of an undergraduate education at U.S colleges and universities) than with traditional on-campus courses and that this difference is greater for courses addressing the academic subject of sustainability (including topics such as globalization, engagement, and citizenship) than for courses unrelated to sustainability.

In sum, a one-size-fits all approach cannot be justified in study abroad and, rather than encouraging students to simply go abroad, academic advisers should attend to the needs of students beyond their desired country of destination and academic goals, to include professional development (including higher-order outcomes such as global citizenry). Incorporating such personal transformational changes into the core objectives of study abroad courses may encourage faculty (and institutions) to acknowledge the added value of study abroad beyond the classroom. It is only in the study abroad environment, for example, that the most dramatic advances in promoting global citizenry are likely to be achieved. Simply put, it is the combination of location (abroad) and academic focus that yields the greatest increases in specified learning outcomes of study abroad.

15.5 Conclusions

In striving to be both innovative and accessible, GPS has been guided by the following objectives: (1) To provide the highest quality, most intellectually and personally challenging and satisfying study abroad experience possible for both students and faculty. (2) To provide programs accessible to a diverse body of students by keeping them as affordable as possible and providing courses suitable for students of all majors and backgrounds. (3) To use the programs as an education framework for developing a body of future scholars and leaders who understand the complex, multi-faceted, global nature of human-environment problems. This reflects our philosophy that sustainable development is not just an issue or problem for scientists or politicians, but requires a well-educated and informed citizenry with a global perspective, sophisticated socio-cultural, economic, and environmental understandings and sense of responsibility and stewardship. (4) To guide students to high personal standards of global citizenship, stewardship and intercultural competence. We believe that students themselves are significantly enriched not only academically but also personally by well-managed international experiences that can have a significant impact on a student's own philosophy and values.

In addition to promoting and enabling high quality education abroad experiences in sustainability for students, GPS has supported faculty participation in education abroad in three distinct ways:

- GPS has evolved a logistical infrastructure that encourages talented faculty—who otherwise might shun the rigors of administering a study abroad trip—to take the leap into leading education abroad. This level of support allows faculty to capitalize on their teaching skills rather than being mired in details such as student recruitment, in-country logistics, or other administration. (Notably, the GPS Faculty Handbook has served as a template for a number of other UGA study abroad programs.)
- GPS offers a curricular infrastructure that provides faculty with a foundation for implementing classes abroad, but still allows for creativity and innovation. The module approach assures even reluctant faculty that they will be offering rigorous academic experiences, yet is sufficiently flexible to accommodate course content ranging from public health to studio arts. Participating faculty members are accorded the autonomy they expect to mount a course within this framework that yet express their unique perspectives on the subject matter.
- GPS prizes the scholarship of teaching and learning and affords faculty with rich opportunities for engaging in associated activities, including detailed student feedback, learning artifacts, and collegial debriefing sessions following each course taught. Such opportunities are consistent with GPS' culture of continuous quality improvement. In addition, the GPS collaborative research program on teaching and learning enables their faculty to publish and present in education abroad.

The programs are guided by a mission statement and set of objectives that seek to position GPS at the forefront of redefining study abroad, as not only a valid academic enterprise, but also as an extraordinary one that can surpass the impact of traditional campus-based instruction. At a time when employers are seeking individuals who are globally connected, with a critical and multi-disciplinary lens to address many of the global issues threatening the sustainability of our planet, research has empirically demonstrated that our students exhibit a significantly broader worldview and sense of global citizenry (relative to pre-departure) than students who do not study abroad or who do not study sustainability while overseas (Tarrant, 2010; Tarrant & Lyons, 2012; Tarrant, Rubin, & Stoner 2014; Tarrant et al., 2011; Wynveen et al., 2012). Moreover, in demonstrating the breadth of impact, GPS provides a first step for many students who otherwise would not consider study abroad. Initial evidence suggests that the opening of this lens is resulting in second (and third) steps, moving UGA beyond the goal of 30 % of graduating students with an international experience to the point where many students can lay claim to an international vocation before graduation.

References

Dolby, N. (2007). Reflections on nation: American undergraduates and education abroad. *Journal of Studies in International Education, 11*(2), 141–156.

Institute of International Education Open Doors Report. (2012). Open doors. http://www.iie.org/en/Research-and-Publications/Open-Doors. Accessed 29 May 2013.

Kegan, R. (2000). What 'form' transforms? A constructive-developmental approach to transformational learning. In J. Mezirow (Ed.), *Learning as transformation: Critical perspectives on a theory in progress* (pp. 35–69). San Francisco: Jossey-Bass.

Mezirow, J. (Ed.). (2000). *Learning as transformation: Critical perspectives on a theory in progress*. San Francisco: Jossey-Bass.

Office of Treaty Settlements. (2008). Central North Island Forests Iwi Collective Deed of Settlement. http://nz01.terabyte.co.nz/ots/DocumentLibrary/CNIsummary.pdf. Accessed 19 Nov 2013.

Perry, L. G., Stoner, K. R., Stoner, L., Wadsworth, D., Page, R., & Tarrant, M. A. (2013). The importance of global citizenship to higher education: The role of short-term study abroad. *British Journal of Education, Society & Behavioral Science, 3*(2), 184–194.

Perry, L., Stoner, L., & Tarrant, M. A. (2012). More than a vacation: Short-term study abroad as a critically reflective, transformative learning experience. *Creative Education, 3*(5), 679–683.

Tarrant, M. A. (2010). A conceptual framework for exploring the role of studies abroad in nurturing global citizenship. *Journal of Studies in International Education, 14*(5), 433–451.

Tarrant, M. A., & Lyons, K. (2012). The effect of short-term educational travel programs on environmental citizenship. *Environmental Education Research, 18*(3), 403–416.

Tarrant, M. A., Lyons, K., Stoner, L., Kyle, G. T., Wearing, S., & Poudyal, N. (2014). Global citizenry and educational travel in Australia or New Zealand. *Journal of Sustainable Tourism, 22*(3), 403–420.

Tarrant, M. A., Rubin, D., & Stoner, L. (2014). The added value of study abroad: Fostering a global citizenry. *Journal of Studies in International Education, 18*(2), 141–161.

Tarrant, M. A., Stoner, L., Borrie, W. T., Kyle, G., Moore, R., & Moore, A. (2011). Educational travel and global citizenship. *Journal of Leisure Research, 43*(3), 403–426.

Wynveen, C., Kyle, G. T., & Tarrant, M. A. (2012). Study abroad experiences and global citizenship: Fostering pro-environmental behavior. *Journal of Studies in International Education, 16*, 334–352.

Chapter 16
Sustainability in Tourism: A Corporate Perspective

Harald Buijtendijk and Mylène van der Donk

Abstract The global tourism industry witnesses significant challenges and changes when it comes to sustainability, changing consumer preferences and ICT developments. Against this backdrop, tourism businesses need to innovate in order to stay relevant for their markets and stakeholders. Critical tourism professionals with strong analytical skills can initiate and guide these processes of innovation in the future, and reach beyond existing blueprints and organizational frameworks. This is the idea behind Sustainability in Tourism—The Corporate Perspective (SUSCOR), a 6-week, English taught, intensive specialization course that explores how sustainable business practices can create (social) returns on investment. SUSCOR is built around a real-life consultancy assignment for a work field commissioner. Students work in international project teams to assess the client's CSR activities in order to identify strategic improvements, design a business case, present this business case to the board of directors at the client's head office, and conclude with a final consultancy report. SUSCOR offers students a platform to practice and develop their problem solving and consultancy skills while simultaneously getting a reality check about their future professional field. As the assignment is embedded in a series of lectures and workshops about contemporary theories about sustainable development, social responsibility and business innovation, SUSCOR links academic knowledge to the reality of today's professional field. This chapter presents the concept and design of the course, illustrates the course week by week, and provides a number of concluding remarks and recommendations that aim to guide the further development of this type of teaching activity.

Keywords Business model innovation • Tourism industry • CSR • License to operate • Stakeholder management • Real-life consultancy assignment

---

H. Buijtendijk (✉) • M. van der Donk
NHTV Breda University of Applied Sciences, Breda, Netherlands
e-mail: buijtendijk.h@nhtv.nl; donk.m@nhtv.nl

G. Moscardo, P. Benckendorff (eds.), *Education for Sustainability in Tourism*, CSR, Sustainability, Ethics & Governance, DOI 10.1007/978-3-662-47470-9_16

16.1 Introduction

16.1.1 Brief Outline of the Teaching Activity

Sustainability in Tourism—The Corporate Perspective (SUSCOR) is a 6-week, English taught, intensive specialization course worth three ECTS (European Transfer System) that is equivalent to a workload of 84 hours. SUSCOR offers students a real-life consultancy assignment for a work field commissioner (their client). In this assignment, students work in international project teams to assess the client's CSR activities in order to identify strategic improvements. Subsequently, they design a business case for one tangible improvement, which they present to the board at the client's head office in the final week of the course. Students are required to compile a report with their findings, which they will offer to their client upon completion of their project. To support the project teams, students are offered four lectures and four workshops (90 min each) over a 5-week period. The course kicks off with a guest lecture by the commissioner, in which the context and purpose of the consultancy assignment is introduced. Students receive a group mark for their report and video summary. An individual normative exam will assess their knowledge of the theory discussed in the course. The commissioner is not involved in the assessment of the students.

16.1.2 Brief Introduction of the Aspect of Sustainability Targeted in the Teaching Activity

All students who participate in SUSCOR have learned about sustainability as an integrated element in the courses of their curriculum. These topics mostly focused on sustainability from the destination and the community perspective. As such they are familiar with various theoretical concepts related to sustainability, they have been trained in identifying positive and negative impacts of tourism on destinations and its stakeholders and have learned about sustainable developments and industry initiatives. This course therefore builds on the already acquired knowledge. In SUSCOR sustainability is approached from a different angle, from the business perspective. SUSCOR explores the application of corporate social responsibility (CSR) as tool for business model innovation in the tourism industry. CSR can be broadly defined as voluntary actions of enterprises to address environmental, social, and economic impacts of their business, in collaboration with their stakeholders (Dahlsrud in van de Mosselaer, van der Duim, & van Wijk, 2012). A business model is the conceptual rather than the financial design of a business. It defines how enterprises deliver value to customers, convince customers to pay for that value, and convert these payments into profit (Teece, 2010). Content wise SUSCOR comprises the accommodation sector, airline industry, and the tour operating industry. However, we have put the main focus in our course on tour operators, because of their considerable influence in directing tourism flows and coordinating supply chains, mainly in mainstream package holiday markets (Budenu, Van Wijk and Persoon in Van de Mosselaer et al., 2012). At the same time, developments in ICT and changing consumer preferences put increasing pressure on their traditional role of travel intermediaries and press the need for innovation in the tour operating industry. Key concepts covered in the course are (corporate) social responsibility (CSR), business model innovation and stakeholder engagement.

16.1.3 Context of the Activity

SUSCOR is designed for students of the International Tourism Management Studies (ITMS) Bachelor program of NHTV Breda University of Applied Sciences. NHTV's vision on education is available online: http://goo.gl/yqiWpq. Within the ITMS program all courses should meet the following criteria to be in line with the graduation profile for BA students:

- **Enhance critical thinking:** Students should be stimulated to perform a critical analysis of a given phenomenon based on basic to profound knowledge of that phenomenon. Students should be able to motivate and explain their conclusions by using relevant examples and/or reliable (academic) sources;
- **Have relevance for the industry:** Lecturers should make use of up to date practical examples and cases, and facilitate students to build their own network with industry organizations through placements, practical assignments and thesis work;
- **Enable students to develop a problem solving approach:** Students should be stimulated to incorporate a problem solving approach in their work. Lecturers should enable this capacity by creating course content, methodology, assessments, and feedback that facilitates students to develop a problem-solving attitude;
- **Incorporate an international attitude:** The ITMS Program is built on the international classroom concept, in which each class (and preferably each group) consists of different nationalities and cultural backgrounds.

The International Tourism Management (ITMS) program consists of two BA specializations, International Tourism & Travel Industry (ITTI) and International Tourism Management & Consultancy (ITMC). Each specialization spans six semesters or 3 years of study. ITTI is predominantly focused on tourism companies along the main tourism supply chain, i.e. tour operators, airlines, and accommodation providers. ITMC takes tourism destinations as a leading perspective, looks at horizontal as well as vertical interrelations in tourism supply chains and networks and concentrates on destination related tourism development challenges. SUSCOR is part of the ITMC program, offered as optional course in the fifth semester. This semester offers ITMC students the possibility to gain more in-depth understanding of a specific topic (minor) prior to starting their individual thesis work in semester 6.

The course participants are 4th year BA students ITMS from various nationalities and cultural backgrounds, including a considerable number of international exchange students from different European universities. The students have obtained basic knowledge about the concepts covered in the course in semester 1, 2 and 3, and gained practical experience—and often a reality check—during their internships in semester 4. The course is located in semester 5 to provide a more in-depth understanding of the subject and to prepare students for their thesis work in their sixth semester. Although students possess the basic knowledge required to be able to follow the course, in-depth understanding concerning issues and challenges surrounding sustainability in a business setting is limited, and students will have to be activated to adopt a business perspective on both sustainability as well as the issues at stake.

16.2 Conceptual Framework of the Activity

16.2.1 Rationale and Development

Conceptually, the course has been based on three premises. First, the supply side driven logic of the industrial era—the idea of seeking profit through growth achieving economies of scale and cost reductions—is no longer viable (Teece, 2010), because it cannot be indefinitely sustained. Second, businesses worldwide have come under increased public scrutiny for causing negative impacts. Therefore the dominant political and economic paradigm of measuring success in strict terms of economic growth is no longer undisputable. Businesses making profit are accepted, but the public wants something back that is tangible and goes beyond increasing stock prices and shareholder value (Blowfield & Murray, 2008). Third, businesses thus need to innovate in order to maintain their *license to operate* in their markets as well as for the general public (ibid). CSR could be used as a mechanism that catalyzes innovation, following the stages of CSR involvement of Zadek (2004). In this process, the business model design logic of Teece (2010) could be used as a tool to reach a stage of engagement, in order to enable tangible innovation.

Aforementioned premises have resulted in a work field that is subject to continuous change, and requires different capacities from our graduates. Future tourism professionals need analytical skills and the ability to think critically in order to be able to come up with new ideas and innovations that go beyond existing blueprints and organizational frameworks. At the same time, these new ideas should be incorporated into sound and logical business concepts that make optimal use of the opportunities new technologies and ICT applications are offering. By offering students a real-life, commissioned consultancy assignment within the parameters of business innovation and CSR they get the opportunity to develop their problem solving and consultancy skills while simultaneously getting a reality check from their future professional field. Furthermore, by embedding this assignment in useful and contemporary theories through lectures and readings, and practical sessions (workshops), SUSCOR contains a strong link between academic knowledge and the reality of the work field.

16.3 Description of the Teaching Activity

16.3.1 Course Synopsis

The course is built around a real-life consultancy assignment for a commissioner (see Appendix 1 for an outline of a possible assignment). The assignment challenges students to take CSR to the highest possible stage of involvement as defined by Zadek (2004) by using CSR as a framework for business model innovation (see Fig. 16.1, Learning Objective 5). Students have to conduct this assignment in groups, and are expected to apply theoretical knowledge as well as insights from cases offered to them in four sets of lectures and workshops. Each set consists of one lecture and one workshop, covers 1 week in the course program, and corresponds with a specific learning objective (see Fig. 16.1). The course starts with an

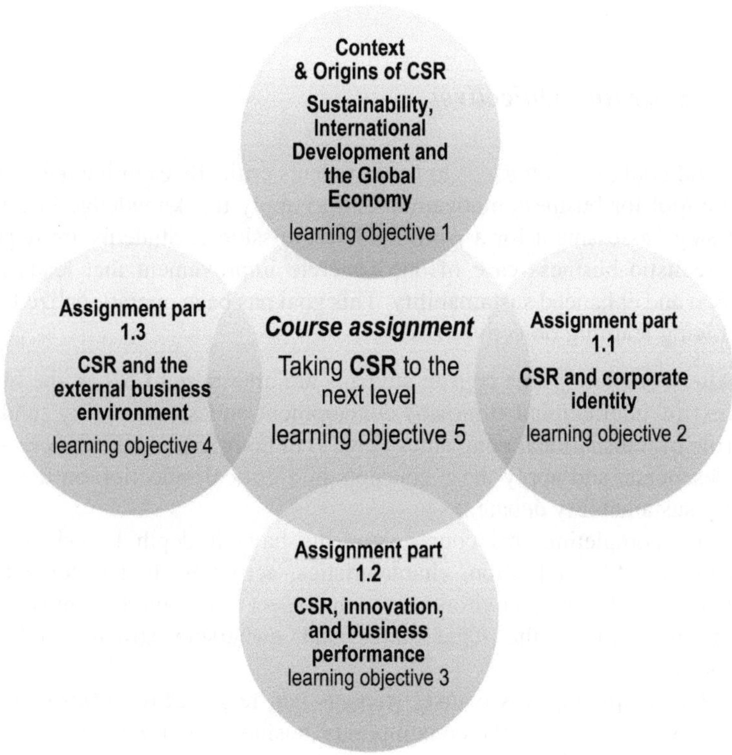

Fig. 16.1 Conceptual model SUSCOR

introduction into the context and origins of CSR (Learning Objective 1), and continues with an introduction of CSR as a concept in relation to the contemporary firm (Learning Objective 2). In week 3 we proceed with linking CSR to business performance and strategic innovation by examining the notion of business models (Learning Objective 3), before concluding the theoretical part of the course with exploring the relation between CSR and the external business environment, looking particularly at stakeholder management.

Throughout the course students are expected to work independently on their assignment alongside the lectures and workshops. The commissioner provides a guest lecture in the first week of the course to introduce the assignment, and will host the students for their final group presentations to the board in the sixth week of the course. Students are required to submit their final group reports in week 7, so that feedback from the presentations can still be incorporated. Each week two plenary sessions of 90 min each are scheduled. These sessions support the preparation of the report; i.e. the focus of each session progressively builds toward the final submission of the assignment. All theory is covered in the first 4 weeks of the course, with each week featuring one lecture and one workshop. In week 5 and 6 only workshops are organized. For inspiration, two sample lesson plans are included in Appendixes 2 and 3.

16.3.2 Learning Objectives

The general goal of the course is to have students critically examine the concept of CSR as a tool for business innovation, and to apply this knowledge in a real life CSR strategy assessment for a work field commissioner. Students are required to build a realistic business case of one concrete improvement that leads towards innovation and enhanced sustainability. This goal has been operationalized through the following learning objectives (LO):

LO1. After completing this course, students are able to put CSR in the historical context of international (tourism) development and sustainability, understand the role of transnational companies (TNCs) in both, explain the concepts power and discourse, and apply these concepts in a critical reflection on the contemporary sustainability debate.

LO2. After completing this course, students have in-depth knowledge of the concept of CSR (definition, characteristics, activities) in the context of the contemporary business environment with respect to corporate identity, mission, vision, as well as the legal status and ownership structure of modern corporations.

LO3. After completing this course, students can relate CSR to business performance in the context of the contemporary business environment as well as the traditional twentieth century business environment, explain key characteristics of both environments, understand and reflect on open and closed innovation and apply the concept of business models in debates about innovation.

LO4. After completing this course, students can relate CSR to the external environment of organizations, indicate the importance of stakeholder management and stakeholder engagement, and discuss the concept *license to operate*.

With respect to LO1–LO4 students are expected to be able to explain all concepts presented in the course by using relevant definitions, critically reflect on their interrelation, and provide relevant examples from the international tourism industry.

LO5. After completing this course, students are able to apply the aforementioned knowledge in a real-life setting and successfully assess the CSR strategy of a tourism company in order to select and identify one tangible business innovation for sustainability presented in a solid business case.

See Appendix 4 for the definition of learning objectives in relationship to modes of assessment and feedback based on the Higher Learning Taxonomy of Dee Fink (2003).

16.3.3 Study Load

The study load of the course is 3 ECTS, or 84 (3×28) h of study. This breaks down in 20 h to attend lectures and workshops, 14 h for exam and workshop preparations, and 50 h for the group work.

16.3.4 Course Resources

Blowfield, M., & Murray, A. (2008). *Corporate responsibility: A critical introduction*. Chapter 1. Oxford: Oxford University Press.

Mowforth, M., & Munt, I. (2009). *Tourism and sustainability: Development, globalisation and new tourism in the third world*. Chapter 1. New York: Routledge.

Teece, D. J. (2010). Business models, strategy, and innovation. *Long Range Planning, 43*, 172–194.

The group work is challenging for students in terms of workload and complexity. Although 4th year students are used to work in groups, working for a commissioner from the industry brings extra pressure. Hence, the emphasis for their final mark is on the group assignment. In order to make sure that theory from the lectures and practical insights from the workshop give optimal support to students with respect to their group work, the content of lectures and workshops has been aligned schedule-wise with the tasks students are performing for their group assignment.

16.3.5 Assessment

Students are graded collectively for their group work (70 % of their final mark) and individually for their normative exam (30 % of their final mark). As the emphasis is on the group assignment (70 % of their final mark), the time allocated to individual study time has been kept to a minimum.

The exam is organized mid-term so that it can serve as a deadline to finish the reading before the practical part of the course has to be completed in week 5–6. Here the focus is on applying the knowledge obtained earlier on in the course. The exam consists of eight open questions, each question worth 12 points (total 96 points so students start with a 4 point bonus if they have completed the course evaluation form at the end of the course). The exam tests factual knowledge (testing students' ability to provide definitions and explain the major concepts covered in the course, with true or false answers) as well as deeper knowledge (testing students' ability to relate concepts and provide solid arguments as well as critical appraisal of the concepts covered in the course). The group work essentially acts as a forward-looking assessment.

The emphasis in testing of both the group work and the exam is on students' ability to make a profound analysis, based on solid arguments and proper motivations. This is reflected in the grading criteria for both the group work and the final exam. In the allocation of points the emphasis is on the argument rather than on the good answer; and particularly rewards students that have succeeded in providing an in-depth analysis of a subject. The process of thinking through the steps of the assignment and making well-argued choices as they move in is essential in this part of their study as it prepares them for their graduation thesis work that starts in semester 6.

The assessment criteria of the normative exam as well as the group assignment can be found in Appendixes 5 and 6.

16.3.6 Course Evaluation

The course is evaluated by asking students to discuss the usefulness of the course in class after completion of the course. Second, all students are required to fill out the course evaluation form. The evaluation form constitutes of quantitative (likert scale scores) and qualitative components (open questions). It provides insights in the perceived quality of the different course components both in terms of content and teaching method. It also evaluates whether the course has been in line with the educational vision of NHTV. Finally, it looks ahead by asking students to suggest improvements for the next edition of the course in the following educational year. Because a thorough evaluation is essential to further improve the quality of this course, students are required to fill out the course evaluation form before handing in their final report (this earns them a bonus of 4 points to be added to their exam mark).

16.3.7 Preparations Prior to the Course

As the course is built around a real-life consultancy assignment for a work field commissioner, it is crucial to collaborate with a company that is interested in acting as official commissioner, and in the position to commit (senior) management level staff to the project. The commissioning company can be either a SME or a Trans National Company (TNC). In the 2012 edition of SUSCOR for instance we collaborated with TUI Nederland NV (www.tui.nl), and in 2013 we partnered with SNP Natuurreizen (www.snp.nl). It is important to approach suitable relations at least 3 months in advance, and set a pre-meeting to discuss the content and purpose of the course. It is necessary to be flexible content-wise and to jointly assess which knowledge needs the company has, before deciding on the exact purpose and scope of the assignment. In terms of time contribution, participation will take the company about 1.5 working days for one (senior) management level staff excluding preparation time. This breaks down as follows: 1 day for the guest lecture, half a day to host the final student presentations. Preferably at least two other management level staff should attend the final presentations. This will add another 0.5 working day per staff member. Finally, it is recommended that the commissioner has staff available that students can consult during the course. In our experience it is most efficient when instructors collect the questions from the students and coordinate all communication between the students and the commissioner in order to avoid double questions and confusion.

Preferably students work together in international teams of 3–4 students. To save time it is recommended to compose the groups in advance. The instructor can handpick them and create balanced groups in terms of student quality, gender, and nationalities. An alternative would be to have students compose their own teams. Additionally, and depending on the amount of time available prior to the course, the instructor could ask students to provide a motivation that explains their team compositions. Students are expected to work according to professional standards and are themselves responsible for managing any disputes or cross-cultural differences in the group. This could be based on a number of basic requirements (i.e. gender, nationality). If difficulties do occur and students consult the instructors, the guideline is that they should take appropriate measures as a group to manage the issue. Instructors can provide advice with respect to the situation but do not intervene; group decisions (as well as the consequences of these decisions) are thus accepted.

16.3.8 Course from Week to Week

In week 1 the course starts with a thought-provoking lecture (L1) on the context and origins of CSR (LO1), students will form groups, and the commissioner will deliver a presentation that introduces the assignment to the students and gives them a clear idea what is expected from them. In that same session, instructors will inform students about the set-up of the course, the method of assessment as well as evaluation. The course continues in week 2 with a lecture on CSR and the contemporary corporate environment (L2, LO2). In the first workshop (WS1) we will follow a participatory approach and take a closer look at the identity of the commissioner's company and brainstorm about key internal and external factors influencing its current and future operations. Week 3 starts with a lecture on CSR, innovation and business performance (L3, LO3). In WS2 students will map the business model of their client, and map drivers and opportunities for innovation. Week 4 features the final lecture on CSR and the external environment (L4, LO4) and concludes with a normative test that covers all theory of the course. Week 5 is an important week. This week starts with an interactive workshop (WS3) in which the different groups share their findings, and rank possible innovations following a participatory method based on jointly defined selection criteria (L05). WS4 features test presentations. Students present their business cases to each other and provide feedback to each other. For this session, students have to prepare video summaries of their business case at the start of their presentation (see Appendix 7 for additional information on the video summary as well as links to sample videos of the 2012 edition of the course). The program in week 6 starts with a final Q&A session in preparation for the final presentations to the commissioner. In the second half of the week students deliver their presentations for the board at the commissioner's office, and collect final feedback from the commissioner on their business case. Finally in week 7 students need to hand in their final report (Table 16.1).

16.4 Evaluation of the Activity

All participating students (20) filled out the evaluation form for the 2012 edition of the course (see Appendix 8 for the evaluation form). With respect to the quantitative evaluation, the course scored an average rating of 4 (on a scale from 0 to 5), which is considered as very good in the context of the Dutch higher education system. The following items scored 4 and higher: clarity of the goal and objectives of the course, the content of L1 and L3, and the workshop in which students could practice their presentations and provide feedback to each other's presentations (WS4). The mid-term normative exam scored below 3.5. Also, students felt the course only helped them somewhat in preparing for their thesis work (score 2.75) as well as strengthen their problem solving skills (score 3.3). When we look at the qualitative evaluation, students enjoyed the quality of the course and perceived it as a course with lots of room for creativity. By far all of them enjoyed the fact that it was a real-life assignment for a company in the work field. This gave them an insight in how things are perceived with a transnational company. They also appreciated the group work as well as the quality of the lectures. As points for improvement students pointed out that the study load is too high, and the exam too

Table 16.1 Course plan

Week	Session	Topic	Reading	Group work
Context and origins of CSR (LO1)				
1	L1	Sustainability, business, and international development	Mowforth and Munt (2009), Chapter 1	Make groups
1	Presentation assignment & Information about the assessment (PI1)	Commissioner will present the group assignment Instructors provide information about the assessment of the course		Start group work
CSR and the corporate environment (LO2)				
2	L2	CSR, corporate identity, mission, vision and the nature and purpose of modern corporations	Blowfield and Murray (2008), Chapter 1	Group work
2	WS1	Corporate identity and business operations		
CSR, innovation and business model development (LO3)				
3	L3	Business environments, innovation and business model development	Teece (2010)	Group work
3	WS2	Business model mapping		
CSR, stakeholder engagement and the external environment (LO4)				
4	L4	CSR, stakeholder engagement and the external environment	Blowfield and Murray (2008), Chapter 5	Article
4	Mid-term exam			Group work
Assess CSR strategy of tourism company and design business case for improvement (LO5)				
5	WS3	Ranking of solutions		Group work
5	WS4	Business case test presentations & feedback		Prepare and show video summaries
6	Q&A session			
6	Presentations for commissioner			
7	Hand in final report			

challenging. Also, they felt the study load has not been equally divided over the course period, with the peak in the last 2 weeks of the course.

16.4.1 Critical Reflection Based on Inputs from the Evaluations

Based on the evaluations we concurred with the students that the workload has been rather high for a 3 ECTS course. We have addressed this issue in the 2013 edition by putting even more emphasis on the workshops (and less on the theoretical part). Also we stressed in the beginning of the course that students should start right away with working on the assignment. Finally, we have changed the deadline for handing in of the final report. In the 2013 edition, students had to hand in the final report in week 6 (during the presentation to the commissioner) instead of week 7. This created a sense of urgency in the group that benefited the interaction between instructors and students and enhanced the quality and depth of the course. Although we have maintained the proportion of the study load for reading and studying literature, we have reduced the number of exam questions and introduced an open book exam. Furthermore, we have reduced the quite rigid and template-style character of the assignment guidelines, leaving more room for creativity and interpretation for the students.

16.5 Conclusion

In this section we provide a number of practical suggestions for those who wish to adapt (parts of) this teaching activities in their courses. First and foremost, it is essential to select an appropriate company to work together with. This organization should be familiar with the university and its graduates, and the management level staff involved should be committed to the project (which means they should be easily available and approachable during the process, and willing to put effort in preparing and hosting the student presentations in their office). Secondly, in working with a commissioner, sound management of expectations is of great importance. It has to be clear up front that the commissioning company will be dealing with students that are primary doing the assignment to learn more about their field of interest and—at the bare minimum—to get their degree and graduate. In this process, the prime concern of the university is the students' interest, and not the company's. At best, the commissioning company will be amazed by the output of the students, but one should consider that as a *nice to have*. Hence the commissioning company should be willing to focus on the process rather than the output of the assignment. Therefore the commissioner should not be directly involved in the assessment process. One can however be indirectly involved; for example by assessing the extent to which students have taken into account the feedback from their commissioner. Finally, we believe it is important that instructors in this course should act as facilitators rather than teachers. It should be made clear from the onset that students are themselves responsible for delivering a professional service to their client. During the process they should manage obstacles themselves. Instructors should guide them by asking questions, focus on the process, and keep a fair amount of distance from the content of the assignment.

To conclude, as lecturers we feel that education about sustainable tourism or sustainable development in general often remains deadlocked in a theoretical and often idealistic discussion that lacks connection with reality. As the importance of sustainability is no longer disputed nowadays, scholars, practitioners and educationalists should put emphasis on how tourism companies (airline companies, the hospitality sector, and intermediaries) can incorporate sustainable business practice in such a way that it creates (social) returns on investment. In this way, students get insight in the challenges of—in this case—a transnational tour operator in relation to sustainability. This experience in our view is extremely valuable, and can only be achieved in a course in which theoretical concepts are directly applied to an actual challenge of a company that they have got to know relatively well during the course.

Appendix 1: Assignment SUSCOR 2012–2013

You work as tourism consultant with CSR Innovators (CSR-I), a reputable international consultancy company specialised in business innovation and corporate social responsibility (CSR). **CSR-I** started operations in 2000 and has since then developed into a global brand, getting CSR policies and strategies up and running for a manifold of brands across the world.

About CSR Innovators

CSR Innovators has established itself as a reputed consultancy firm with a client portfolio featuring major transnational corporations including global brands such as Boeing, Exxon Mobile, Motorola, Shell, and Unilever. Over the past decade, CSR-I established a solid reputation across sectors; the company won high profile, multi-million dollar contracts in, among others, the construction business, the aviation industry, petrochemicals, telecommunications, logging and mining. In 2004 it launched **CSR Hospitality Innovators**, a separate business unit specifically focusing on the hospitality, tour operating, and transport business. From that year onwards, Hospitality Innovators has gradually established itself as *the* market leader in CSR advisory services for the global tourism industry, with clients including major airline companies, destination marketing organizations, tour operators, and international hotel chains.

The CSR-I Approach

CSR-I takes the strengths and assets of their clients as a starting point, reframing policies and strategies from within, without losing grips with the complex external environment of the global market place. Its focus is on the DNA of companies; CSR-I aims to rework a company's business model in order to secure long term operational sustainability based on a triple bottom line (people—planet—profit). Through this truly innovative approach it has taken CSR from boardroom discussions to everyday business operations! In order to enhance creativity of CSR-I projects, it is company policy to employ experts in special project teams on specific assignments. During the process, there are regular meetings with other project teams working on similar or complementary assignments, in order to create synergy through exchange of knowledge and ideas, and make maximum use of creativity.

Background

Small and medium sized tourism enterprises catering to niche markets in Europe are confronted with consolidating growth rates and decreasing profit margins, which puts budgets under pressure. At the same time, change in the global business arena (new competitors, changing consumer preferences, and demographic trends, and the rapid introduction of new technologies) requires all market players to invest in innovation in order to stay relevant in the decades to come. As a result of such uncertainties, these companies cut in their CSR related spending: particular if the return on investment of CSR activities is low and uncertain.

Your Assignment

As part of their strategy to secure the long-term future of the company, [Name company commissioner] is developing a new brand, and has requested CSR-Hospitality Innovators to assess which social responsible product features should be embedded in this brand and how this should be done.

The company's director, [Name], will act as your commissioner.

The commissioner is willing to take your recommendations in consideration, provided **there is a solid business case that demonstrates how the suggested action(s) enhance(s) the company's overall business performance within 3 years**.

The management of CSR Hospitality Innovators has appointed two project teams to this client that will work simultaneously on this assignment, to facilitate maximum creativity. In your capacity as CSR consultant, you will be part of one of these teams.

In a time span of 6 weeks, your team is asked to do two things:

(A) **Assessment of the client's company and brand identity, business performance and external business environment (deliverable I)**
Your team should analyse the client's current CSR strategy with respect to the strategic pillar selected in relation to its corporate identity (1), its business environment (2), and business performance (3). Teams working on the same pillar should differentiate their focus in coordination with each other.

(B) **Build a business case that presents one tangible improvement with regard to this strategic pillar (deliverable II)**
Based on the assessment your team should identify improvements, select the best improvement (1!) and present this improvement in a solid business case. This business case should demonstrate the return on investment the improvement will generate (1) and its contribution to the company's overall business performance (2).

Both the assessment and the business case should be put in a final report and handed over to client in week 7. A template and guidelines for the final report can be found below. Prior to handing over the final reports, your team will deliver a presentation of the major findings to the commissioner in their head office on **October 10th** (Week 6). Specific guidelines for this presentation are as well below.

Work Plan

Week 1

Putting together the project teams. Your team starts to work on part I of the assignment.

Week 2 & 3

Each team will work on part I of the assignment and assess the commissioner's CSR strategy with respect to the selected pillar, its corporate identity, business environment and business performance, and formulates preliminary recommendations for improvements. In week 3, your team prepares a **one page summary** based on your findings so far, for the meeting in week 4.

Week 4 & 5

The three project teams have a plenary brainstorm meeting in which each team present its mid-term findings based on their one page summaries. The aim is to exchange ideas and discuss possible improvements to the current CSR strategy of the commissioner clients. Based on the first scan (week 2) and this brainstorm meeting (week 3), each team will now perform the final analysis of the client's CSR strategy, select the best improvement (1!) based on prior findings and inputs from the brainstorm meeting, and build the business case for this improvement. Each team will prepare a

- presentation for the commissioner, in which it will summarize major findings, and
- present their business case. As a prelude to this presentation (and summary of the project), your team will prepare a **2 minute video clip** (see Appendix 2) in which you will summarize your major findings for the commissioner.

Week 6

The week starts with a preparations in which you can test and discuss your presentation with the other teams. Subsequently, presentations to the commissioner will take place on **Wednesday, October 10th** from 12.30 to 16.00 in the head office of the commissioner. Based on the feedback of the commissioner, you can start finalizing your final reports.

Week 7

Complete and hand in final report.

Appendix 2: Sample Lesson Plan Lecture 3 (Week 3)

	L3 (week 3)
Title	**The world is upside down** **CSR & innovation in a changing business environment**
Learning Objective covered	After completing this course, students can relate CSR to business performance in the context of the contemporary business environment as well as the traditional twentieth century business environment, explain key characteristics of both environments, understand and reflect on open and closed innovation and apply the concept of business models in debates about innovation (LO3)
Course reading	Teece (2010). Business models, strategy, and innovation. Long Range Planning 43 (172–194)
Total time	**120 min**

Specifications

Time	Topic	Purpose	Contents	Teaching method & materials	Student activity	Teacher activity
10 min	**Introduction: Positioning this lecture**	Looking back, looking ahead Test prior knowledge Provide student with the context of today's lecture Position the lecture in the course	Look back at previous class and ahead to the coming weeks Highlight LO3 Introduce concept *innovation* Link concept innovation to CSR	Plenary presentation creating room for interaction PPT	Ask questions and answer questions	Lecturing while creating room for interaction by regularly asking questions to students
15 min	**PART 1: Traditional business environment**	To explain the origins and characteristics of the traditional twentieth century business environment Providing a historic overview	Industrial Era Origin of traditional production Rationale of traditional production: economies of scale Benefits Passive role of consumer One dimensional sales & marketing Examples	Plenary presentation creating room for interaction PPT & Youtube video	Ask questions and answer questions, small discussion	Lecturing while creating room for interaction by regularly asking questions to students
15 min	**PART 2: Contemporary business environment**	To explain factors that catalysed change from traditional to contemporary business environment To explain the origins and characteristics of the contemporary business environment	Rise of the new web Increased consumer power Wisdom of crowds User generated content Social production vs. mass production	Plenary presentation creating room for interaction PPT & Youtube video Smartphone experiment in class	Ask questions and answer questions, small discussion	Lecturing while creating room for interaction by regularly asking questions to students
20 min	**PART 3: Group exercise**	How do changes in the business environment impact the company students are analysing in the SUSCOR group assignment		Group work Markers & flipchart papers	Students work in the same group as the course assignment and list possible changes that affect their client's company and motivate their findings. They list their findings on	Walk around. Guiding by asking questions Wrap up first half of lecture after discussion

(continued)

Specifications						
Time	Topic	Purpose	Contents	Teaching method & materials	Student activity	Teacher activity
					flipchart paper and briefly present to each other Discussion during presentations	
20 min	**PART 4: Business model development**	Explain what a business model is and how business model development can be used as a tool for business innovation	Explain concept and origins Characteristics of business models Purpose of business model development Link this back to CSR	Plenary Presentation creating room for interaction PPT & Youtube video	Ask questions and answer questions, small discussion	Lecturing while creating room for interaction by regularly asking questions to students
20 min	**PART 5: Example video and Group exercise**	Watch a video that explains the business model innovation of KLM Identify the 3 steps of the business model in the company they are analysing for the SUCOR group assignment and motivate their findings		Group work Markers & flipchart papers	Students work in the same group as the course assignment and list possible changes that affect their client's company and motivate their findings. They list their findings on flipchart paper and briefly present to each other Discussion during presentations	Walk around. Guiding by asking questions Wrap up second half of lecture after discussion
10 min	**Final wrap up**	Reflection: what did the students gain? Looking forward	Summary Looking to next week	Plenary	Ask questions and answer questions, discussion	Ask questions

Appendix 3: Sample Lesson Plan Workshop 3 (Week 5)

Lecture	**WS3 (week 5)**
Title	**Assessment and ranking of solutions for the business case**
Learning Objective covered	After completing this course, students are able to apply the aforementioned knowledge in a real-life setting and successfully assess the CSR strategy of a tourism company in order to select and identify one tangible business innovation for sustainability presented in a solid business case (LO5)
Student preparation	Groups prepare one page summary of their findings and indicate two possible improvements for the CSR strategy of their client
Total time	**100–120 min**

Specifications

Time	Topic	Purpose	Contents	Teaching method & materials	Student activity	Teacher activity
10 min	**Introduction: Going through the set up of the session**	Lay out purpose and relevance and activities of the session	To discuss issues/challenges related to implementing CSR To identify solutions for these challenges To rank these solutions according to preset criteria (business case)	Plenary presentation creating room for interaction	Ask questions and answer questions	Explaining and checking whether everything is clear
15 min	**SESSION I Statements**	Learning from each other	Group share findings Groups state/ challenges issues and motivate why these issues/ challenges are important	Group discussions and presentation Whiteboard and markers	Each group briefly outlines their own findings and states their issues/ challenges Each group explains *why* these issues/ challenges are important	Facilitator puts issues/challenges on the whiteboard and make an overview
15 min	**Reading and discussion time**	Learning from each other	Groups exchange their one-page summaries and discuss work of other groups	Reading/ sharing and Group discussions Post it notes Whiteboard and markers	Distribution of summaries (each group will read the work of one of the other groups). Each group will	Facilitators puts post its on the whiteboard at the appropriate spot

(continued)

Specifications						
Time	Topic	Purpose	Contents	Teaching method & materials	Student activity	Teacher activity
					discuss summary in the group in order to identify possible solutions to tackle these issues/ challenges. Each groups puts solutions on post it notes	
20 min	**SESSION II Discussion of possible solutions**	Learning from each other Discussion to find the best way forward	Selection of best possible solution Group presents these solutions to other groups Explain why these solutions have been selected Identify obstacles that could prevent these solutions from being implemented	Group discussion and presentation Post it notes Whiteboard and markers Flipchart papers	Discussion and decision in group which solution is most appropriate Each group will present their solutions, explain why these are solutions are appropriate, and what could be obstacles preventing these solutions from being implemented	Distribution of roles individual group members (each member will focus on specific section; there are 3 different sections corresponding with part 1.1, 1.2 and 1.3 of the group assignment. Guiding by asking questions
10 min	**INTRODUCTION OF RANKING GRIT**	To explain the ranking grit to students	Ranking grit is 6 criteria of business case in template	Plenary Hand out Ranking grit	Ask questions	Plenary explanation
25 min	**SESSION III SELECTION OF BEST SOLUTION**	To select best solution based on criteria for a good business case	Apply each business case criteria to possible solutions Select the best solution	Group discussion and presentation Post it notes Whiteboard and markers Flipchart papers	Each group ranks their own solutions according to the ranking grit on a white paper. Presenting these to the other groups	Guide by asking questions, facilitate presentations and discussion
10 min	**Final wrap up**	Reflection: what did the students gain? Looking forward	Summary Looking to next week	Plenary	Ask questions and answer questions, discussion	Ask questions

Appendix 4: Definition of Learning Objectives in Relationship to Modes of Assessment and Feedback

Learning Objective 1

After completing this course, students are able to put CSR in the historical context of international (tourism) development and sustainability, understand the role of transnational companies (TNCs) in both, explain the concepts power and discourse, and apply these concepts in a critical reflection on the contemporary sustainability debate.

LO1	Foundational knowledge	Integration & evaluation	Application	Value & care
Specification	Provide a basic definition of CSR Explain International Development based on the Truman Doctrine Name limitations of this perspective on Development Provide a basic definition of Sustainability List characteristics of multinational companies Explain what Dependency theory is about Explain what Neoliberal economics is about Explain the concepts power and discourse	Appraise the relationship between international development and sustainability using dependency theory and Neoliberal Economics Appraise the impact of TNCs on developing countries using Dependency theory Appraise the benefits TNCs bring to developing countries using Neoliberal Economics Discuss the opportunities and limitations for transnational companies to contribute to sustainability		Student appreciation of the content covered in week 1 of the course Student appreciation of the learning methods applied in week 1 of the course
Mode of assessment	Normative test week 4	Normative test week 4		Evaluation meeting with students after completion of the course
Feedback mode & moment	Feed forward in PI1 in week 1 Evaluation meeting with	Feed forward in PI1 in week 1 Exam review session 10 days after		Evaluation meeting with students after completion of the course

(continued)

LO1	Foundational knowledge	Integration & evaluation	Application	Value & care
	students after completion of the course	examination Evaluation meeting with students after completion of the course		

Learning Objective 2

After completing this course, students have in-depth knowledge of the concept of CSR (definition, characteristics, activities) in the context of the contemporary business environment with respect to corporate identity, mission, vision, as well as the legal status and ownership structure of modern corporations.

LO2	Foundational knowledge	Integration & evaluation	Application	Value & care
Specification	Provide a basic definition of CSR List components of CSR List different CSR activities Explain motivations for companies to engage with CSR Provide basic definitions of corporate identity, mission & vision Provide a basic definition of a corporation List legal characteristics of corporations	Evaluate the motivation of corporations to engage in different CSR activities based on the purpose and nature of corporations Appraise conflicting interests of corporations with respect to CSR engagement in relation to their legal status	Assess the CSR strategy of your client focusing on corporate identity, mission, vision, and current CSR strategy (Assignment part 1.1)	Student appreciation of the content covered in week 2 of the course Student appreciation of the learning methods applied in week 2 of the course Student appreciation of the group assignment of the course as a whole as well as part 1.1 in particular
Mode of assessment	Normative test week 4	Normative test week 4	Group assignment	Evaluation meeting with students after completion of the course
Feedback mode & moment	Feed forward in PI1 in week 1	Feed forward in PI1 in week 1 Exam review	Peer feedback during WS1 in week 2 Q&A session in	Evaluation meeting with students

(continued)

LO2	Foundational knowledge	Integration & evaluation	Application	Value & care
	Evaluation meeting with students after completion of the course	session 10 days after examination Evaluation meeting with students after completion of the course	week 6 Expert feedback during final presentation in week 6 Feedback on final result 10 days after handing in the final report based on assignment assessment form	after completion of the course

Learning Objective 3

After completing this course, students can relate CSR to business performance in the context of the contemporary business environment as well as the traditional twentieth century business environment, explain key characteristics of both environments, understand and reflect on open and closed innovation and apply the concept of business models in debates about innovation.

LO3	Foundational knowledge	Integration & evaluation	Application	Value & care
Specification	Provide a basic definition of innovation Understand the difference between open and closed innovation List characteristics of the traditional twentieth century business environment List characteristics of the contemporary business environment Provide a basic definition of business model development List the main	Debate the challenges as well as opportunities of the rapidly changing business environment for tourism companies using the theory of Teece (2010) on business model development Appraise these challenges and opportunities in relation to the CSR strategy of these companies using the theory of Teece (2010) on business model development	Assess the CSR strategy of your client focusing on business performance applying the main steps of business model development (Assignment part 1.2)	Student appreciation of the content covered in week 2 of the course Student appreciation of the learning methods applied in week 2 of the course Student appreciation of the group assignment of the course as a whole as well as part 1.2 in particular

(continued)

LO3	Foundational knowledge	Integration & evaluation	Application	Value & care
	steps of business model development Understand the purpose of business model development			
Mode of assessment	Normative test week 4	Normative test week 4	Group assignment	Evaluation meeting with students after completion of the course
Feedback mode & moment	Feed forward in PI1 in week 1 Evaluation meeting with students after completion of the course	Feed forward in PI1 in week 1 Exam review session 10 days after examination Evaluation meeting with students after completion of the course	Peer feedback during WS2 in week 3 Q&A session week 6 Expert feedback during final presentation in week 6 Feedback on final result 10 days after handing in the final report based on assignment assessment form	Evaluation meeting with students after completion of the course

Learning Objective 4

After completing this course, students can relate CSR to the external environment of organizations, indicate the importance of stakeholder management and stakeholder engagement, and discuss the concept *license to operate*.

LO4	Foundational knowledge	Integration & evaluation	Application	Value & care
Specification	List the main stakeholders of the external environment of these organizations List the main external treats and opportunities of an organization Explain the	Discuss external treats for large tour operators using the concept licence to operate Use the concept licence to operate to appraise how these organizations could use	Assess the CSR strategy of your client focusing on the external environment, stakeholders, and stakeholder management and engagement	Student appreciation of the content covered in week 3 of the course Student appreciation of the learning methods applied in week 3 of the

(continued)

LO4	Foundational knowledge	Integration & evaluation	Application	Value & care
	difference between stakeholder management and stakeholder engagement List reasons why stakeholder management and stakeholder engagement is important for organizations Provide a basic definition of the concept licence to operate	stakeholder engagement in business innovation for sustainability	(assignment part 1.3)	course Student appreciation of the group assignment of the course as a whole as well as part 1.3 in particular
Mode of assessment	Normative test week 4	Normative test week 4	Group assignment	Evaluation meeting with students after completion of the course
Feedback mode & moment	Plenary in week 4	Feed forward in PI1 in week 1 Exam review session 10 days after examination Evaluation meeting with students after completion of the course	Peer feedback during WS3 in week 5 Q&A session in week 6 Expert feedback during final presentation in week 6 Feedback on final result 10 days after handing in the final report based on assignment assessment form	Evaluation meeting with students after completion of the course

Learning Objective 5

After completing this course, students are able to apply the aforementioned knowledge in a real-life setting and successfully assess the CSR strategy of a tourism company in order to select and identify one tangible business innovation for sustainability presented in a solid business case.

LO5	Foundational knowledge	Integration & evaluation	Application	Value & care
Specification			Assess the main findings of the CSR strategy analysis of LO1, LO2, LO3 and LO4 in order to identify one tangible improvement for the client based on solid indicators defined by the group Motivate the choice for these indicators as well as the selection results Design a solid business case that demonstrates the value of the improvement to the client	Student appreciation of the content covered in week 4, 5 and 6 of the course Student appreciation of the learning methods applied in week 4, 5 and 6 of the course Student appreciation of the group assignment of the course as a whole as well as part 2 in particular
Mode of assessment			Group assignment	Evaluation meeting with students after completion of the course
Feedback mode & moment			Peer feedback during WS3 in week 5 Q&A session in week 6 Expert feedback during final presentation in week 6 Feedback on final result 10 days after handing in the final report	Evaluation meeting with students after completion of the course

Appendix 5: Assessment Criteria Normative Final Exam

Type of question	True/false questions (A)	Argumentation (B)
Answer required	Name any of the following: Characteristics, benefits, constraints, yes/no answers, examples, activities, threats, reasons (argument in favour or against) Or correct/incorrect use of concept in answers	Explanation, illustration, appraisal of theory, motivation
Maximum scores	Q1 (4), Q2 (4), Q3 (5), Q4 (6), Q5 (4), Q6 (4), Q7 (4), Q8 (4)	Q1 (8), Q2 (8), Q3 (7), Q4 (6), Q5 (8), Q6 (8), Q7 (8), Q8 (8)

(continued)

Type of question	True/false questions (A)	Argumentation (B)
obtainable per question		
Maximum score (total 96 points)	35 points	61 points
Grading criteria	Incorrect: 0 % score Correct: 100 % score	0 % score Answer is incorrect/incomprehensive/lacking 50 % score One or more of the following deficiencies has occurred: Answer is incomplete with respect to what is asked Answer is too general and/or required in depth analysis and/or critical reflection is lacking Answer has no coherent argumentation Answer is partly correct/partly incorrect 100 % score No deficiencies have occurred: Answer is correct/complete/has sufficient depth in terms of analysis and or critical reflection/has coherent argumentation

Appendix 6: Assessment Criteria and Feedback Form Group Assignment

Break down of your score

Criteria	Max. score	Complete y/n based on template assignment	Quality
lay-out (bonus)	5		0 % far below standards 40 % acceptable 60 % professional 100 % of exceptional quality
Referencing (bonus)	5	Y = 100 % N = 0 %	n/a
Video Summary	10		0 % far below standards 40 % acceptable 60 % professional 100 % of exceptional quality

(continued)

Criteria	Max. score	Complete y/n based on template assignment	Quality
Part 1 section 1	15	n/a	Too general description AND/OR attention for irrelevant details or content = 20–60 % max To the point, relevant, and with authority = 60–100 %
Part 1 section 2	15	n/a	Too general description AND/OR attention for irrelevant details or content = 20–60 % max To the point, relevant, and with authority = 60–100 %
Part 1 section 3	20	n/a	Too general description AND/OR attention for irrelevant details or content = 10–50 % max To the point, relevant, and with authority = 60–100 %
Results of the assessment	20	n/a	Too general description AND/OR attention for irrelevant details or content = 10–50 % max To the point, relevant, and with authority = 60–100 %
Business case	20	Element missing = −20 %	Too general description AND/OR attention for irrelevant details or content = 10–50 % max To the point, relevant, and with authority = 60–100 %
Sub total			
Total score	110		

Appendix 7: Executive Summary Videoclip

Instead of a written executive summary, students are asked to include an executive summary video clip in their report. The advantage of such a video clip is that it sparks creativity among the students and gives students something they can share with the commissioner prior to their final presentation. In this way, the commissioner can get an idea of what the different business cases are about without spending too much time with going through draft versions of student reports.

See the links below for the videos of the 2012 edition of the course:

- http://goo.gl/mgtACb
- http://goo.gl/edtc6j
- http://goo.gl/xKAu01
- http://goo.gl/uQsczd

Students were provided with the following instructions with respect to the video clip:

1. **Content**
   Your video must include the following items:

- Brief introduction of your project (state the objective and relevance of your research);
- Briefly explain your way of work;
- Most important finding;
- Which improvement have you selected and why;
- The business case of this improvement (see part II of assignment).

2. **Video production**

- Length video: between 2 and 3 min. No longer than 3 min!
- Be original and engaging; try to film at different locations. Can be both inside and outside NHTV. Feel free to add audio or visual effects;
- Audience your message well. Keep in mind you are aiming your message to management level professionals;
- Make sure the message is clear (see content). The aim is to give the commissioner a clear idea of the work you did and what you have come up with, so that they can prepare themselves for your presentation;
- Use English language only.

3. **Video editing**
   Often smartphones come with basic video editing software. Free video editing software is available online. This for instance is a handy one for Windows: http://www.nchsoftware.com/videopad/tutorial.html Macs have their own video-editing software.
4. **Hand in**

- Hand in your final video latest on dd/mm/yy, before (time) by sending the final file (format mp4 or mov) by We transfer; (https://www.wetransfer.com/) to instructor@suscor.com
- We will share the video via YouTube (viewed with link only) with you and the commissioner;
- Include the URL to your video in the final report

Appendix 8: Course Evaluation Form SUSCOR 2012–2013

Please indicate below how you perceived the quality of this course by ranking your appreciation on a five point scale ranging from very low quality/very weak/strongly disagree (1) to very high quality/very strong/strongly agree (5).

Please be frank in your answers; socially desired answers do not help us improving this course for the next year! Thank you for your comments!

General quality	**1**	**2**	**3**	**4**	**5**
1. Clarity of the goal and objectives of the course	O	O	O	O	O
2. General content (topics covered in class)	O	O	O	O	O
3. Study materials (articles, book chapters)	O	O	O	O	O
4. Teaching methods: lectures	O	O	O	O	O
5. Teaching methods: workshops	O	O	O	O	O
6. Group Assignment	O	O	O	O	O
7. Exam	O	O	O	O	O
8. Quality of feedback you received during the course	O	O	O	O	O

Remarks:

The course from week to week	**1**	**2**	**3**	**4**	**5**
WEEK 1					
9. Lecture context & origins CSR					
Content (relevant topics for the course)	O	O	O	O	O
Quality of the lecture	O	O	O	O	O
10. Guest lecture group assignment by the commissioner week 1	O	O	O	O	O
WEEK 2					
11. Lecture introduction to CSR					
Content (relevant topics for the course)	O	O	O	O	O
Quality of the lecture	O	O	O	O	O
12. Workshop shared values (organization & target group					
Content (relevant topics for the course)	O	O	O	O	O
Quality of the workshop	O	O	O	O	O
WEEK 3					
13. Lecture business environments, innovation and business model development					
Content (relevant topics for the course)	O	O	O	O	O
Quality of the lecture	O	O	O	O	O
14. Workshop business model mapping					
Content (relevant topics for the course)	O	O	O	O	O
Quality of the workshop	O	O	O	O	O
WEEK 4					
15. Lecture stakeholder engagement					
Content (relevant topic for the course)	O	O	O	O	O
Quality of the lecture	O	O	O	O	O
WEEK 5					
16. **Workshop business case design**					
Content (relevant topic for the course)	O	O	O	O	O
Quality of the workshop	O	O	O	O	O

WEEK 6					
17. **Workshop: test presentations**	O	O	O	O	O
18. **Final presentations head office commissioner**	O	O	O	O	O

Remarks:

Relationship with NHTV Education criteria	**1**	**2**	**3**	**4**	**5**
19. This course helped me to prepare for my thesis work	O	O	O	O	O
20. This course helped me to make a more in depth analysis of a subject	O	O	O	O	O
21. This course improved my problem solving skills	O	O	O	O	O
22. This course helped me to think more critically about a subject	O	O	O	O	O

Write down one or two aspect of the course you really enjoyed:

Write down one or two aspects you think we should improve:

To conclude, please add your general opinion and suggestions.

Thank you for your feedback!

References

Blowfield, M., & Murray, A. (2008). *Corporate responsibility: A critical introduction*. Oxford: Oxford University Press.

Dee Fink, L. (2003). *Creating significant learning experiences: An integrative approach to designing college courses*. San Francisco: Wiley.

Mowforth, M., & Munt, I. (2009). *Tourism and sustainability: Development, globalisation and new tourism in the third world*. New York: Routledge.

Teece, D. J. (2010). Business models, strategy, and innovation. *Long Range Planning, 43*(2–3), 172–194.

Van de Mosselaer, F., van der Duim, R., & van Wijk, J. (2012). Corporate social responsibility in the tour operating industry: The case of Dutch outbound tour operators. In D. Leslie (Ed.), *Tourism enterprises and the sustainability agenda across Europe* (pp. 71–92). Farnham: Ashgate.

Zadek, S. (2004). Paths to corporate responsibility. *Harvard Business Review, 82*(12), 125–132.

Chapter 17
Education for Sustainability Futures

Pierre Benckendorff and Gianna Moscardo

Abstract This chapter concludes the book by considering the main themes and key conclusions that can be drawn from the various chapters that have been included. The chapter also acknowledges some of the major gaps in the book by providing some initial thoughts about key stakeholders and dimensions of sustainability that have not been discussed in other chapters. These include EfS in technical and vocational education sectors, government officials and NGOs. The need for innovation in curricula, pedagogies and technology enhanced learning are explored as key dimensions of EfS. The chapter concludes with a discussion of the future for both sustainability in tourism and EfS.

Keywords Education for sustainability • Tourism education • Tourism and sustainability

17.1 Introduction

The single most dominant theme that underpins all discussions about sustainability and education is the need for profound and fundamental change in the way we think about the world. This change will not happen without a similar change in the way we think about education. According to Carp (2012, p. 223) "*academic knowledge practices are complicit in creating, justifying, maintaining, and applying the behavior that place us at risk*". Arguably this is true both in general and for the tourism academy. Many recent critiques of tourism conclude that despite considerable rhetoric about sustainability the practice of tourism is becoming less sustainable using phrases such as "*far from sustainable*" (Buckley, 2012, p. 534), "*tourism is becoming less sustainable*" (Gössling, Hall, Ekström, Engeset, & Aall, 2012, p. 899), and that the steps that have been taken in tackling tourism and sustainability "*are not remotely enough*" (Bramwell & Lane, 2013, p. 1). There is a growing

P. Benckendorff (✉)
The University of Queensland, Brisbane, QLD, Australia
e-mail: p.benckendorff@uq.edu.au

G. Moscardo
James Cook University, Townsville, QLD, Australia
e-mail: Gianna.moscardo@jcu.edu.au

G. Moscardo, P. Benckendorff (eds.), *Education for Sustainability in Tourism*, CSR, Sustainability, Ethics & Governance, DOI 10.1007/978-3-662-47470-9_17

movement to seriously reconsider our approach to both tourism practice and education and this edited volume is a result of that movement.

The aim of this chapter to bring this edited volume to close. It will do this by considering three main areas: the main themes and key conclusions that can be drawn from the various chapters that have been written for the book, the things that are missing from the book, and a consideration of the future for both sustainability in tourism and EfS to support this.

17.2 Education for Sustainability in Tourism: Key Dimensions and Themes

Rather than adopting a narrow focus on education in formal institutional settings, this edited volume has highlighted the importance of Education for Sustainability (EfS) in tourism across a range of stakeholder groups and settings. Chapter 1 argued that educational systems focused on competition and consumption are moving learners away from sustainability. Instead, it was argued that EfS requires fundamental changes in the nature of education and a focus on education as a transformative experience for learners. EfS in tourism should be characterized by the following features:

- It is transformative both for individuals and groups;
- It is driven by and supports intrinsic values linked to human and environmental well-being;
- It is aimed at capacity building;
- It goes beyond the formal classroom by engaging with multiple stakeholders to tackle complex multi-disciplinary problems;
- It takes a whole system approach; and
- It follows the core principles of sustainability.

The extension of learning beyond the classroom removes the constraints of time and place and recognizes the need to engage with a range of stakeholders through public education campaigns, social marketing, training for decision makers in government and business, and support for community learning to reach people outside the formal educational system. The key stakeholders in EfS in tourism are identified as formal educational institutions, destination communities and tourism organizations. The development of effective EfS in tourism requires that these stakeholders be engaged both as learners and educators.

Part II of this volume provided a foundation for understanding core concepts in tourism EfS. Chapter 2 (Moscardo) identifies a number of challenges, conflicts and core knowledge areas that elsewhere have been referred to as ‘threshold concepts’ and ‘troublesome knowledge’. Fennell’s overview of ethics in Chap. 3 reminds us of the foundations of ethics knowledge but cautions that there is often a disconnect between ethical knowledge and behavior. Like business ethics, tourism ethics suffers from a tension between the need to remain competitive while at the same time maintaining high ethical standards. Attention is again drawn to a concern that the current education system values competition and consumption over caring and conservation.

In Chap. 4, Lund-Durlacher extends the discussion of ethics to corporate social responsibility (CSR). CSR provides a particularly useful framework for helping businesses and destination communities to integrate social and environmental concerns into their operations, strategies and interactions with stakeholders. Examples of key CSR strategies include environmental protection, fair working conditions for employees and contributing to the welfare of local communities. However, CSR can only be achieved through effective and coherent governance systems. In Chap. 5 Dredge notes that tourism professionals increasingly work in complex settings that require them to engage in dynamic social discourses where difficult trade-offs are made between competing demands. She proposes a philosophic tourism practitioner education consisting of three key dimensions: core knowledge for governance, competencies in a range of areas, and ethical action-oriented practice. Educators are challenged to create opportunities that allow learners to reflect on their agency and their ability to enact more sustainable tourism futures.

Part III examined issues and processes relevant to understanding tourism and sustainability in the formal educational sector, including universities, vocational training and school settings. In Chap. 6, Lund-Durlacher describes how sustainability was integrated into a higher education institution by transforming the university's strategy, organizational structure, curricula and teaching strategies. The key lesson from this example is that tourism EfS needs to go beyond knowledge to include active and participatory learning processes because learning also takes place through the 'hidden' curriculum. Educators act as role models and learners are inspired and motivated through sustainability in action. In Chap. 7, Wearing and colleagues examine the assessment and feedback mechanisms that reinforce a values-based approach to tourism EfS. The authors advocate experiential learning and assessment approaches that practically engage learners in a reflective critical analysis of their learning about sustainable tourism. In Chap. 8, Whitehouse examines efforts to embed EfS into the early childhood, primary school and secondary school sectors using Australia as a case study. Like previous chapters in this section, there is a particular emphasis on sustainable practice, with educational partnerships between schools and the tourism industry seen as an enabler of a more sustainable future.

In **Part IV** several chapters explore learning and sustainable tourism in non-institutional settings, including destination communities, business coaching and mentoring and visitor learning. Moscardo and Murphy (Chap. 9) examine the role of EfS in the development of various types of capital that support destination community well being. In particular, this chapter highlights the need for education *about* tourism rather than *for* tourism so that destination communities can have a more meaningful debate about preferred tourism development trajectories and futures. The authors note that education alone is not sufficient to improve sustainability. Destination communities also require opportunities for citizen involvement in tourism governance. Action research projects that facilitate interactions between local communities and learners offer one approach to creating the capital needed for community involvement in tourism governance. In Chap. 10 Blackman and Bauld explore the role of business coaching as a life-long learning strategy to support micro and small enterprises (MSEs) to shift toward more sustainable thinking, culture and practice. Visitors represent another stakeholder group that may be engaged through EfS in tourism. Moscardo's contribution in Chap. 11 brings together the well-established literature on interpretation and visitor learning with developments in the area of sustainability marketing. She argues that developments in marketing may contribute to EfS if ideological differences between marketers and interpreters can be reconciled.

Part V offered a collection of cases to illustrate the use of different pedagogies and assessment approaches in education for sustainability in tourism. Jennings and Kachel ("Online Learning: Reflections on the Effectiveness of Undergraduate Sustainability Tourism Module") use action research to evaluate the effectiveness of an online module focusing on education *for* sustainability as well as education *about* sustainability. Learners were required to demonstrate higher order thinking through a series of online activities, including a reflexive journal. In Chap. 13, Wilson discusses how critically reflexive practice (CRP), as part of a critical theory pedagogy, can be used to transform the teaching experience by motivating educators and providing a stronger nexus between teaching and research in sustainable tourism. Using the three cornerstones of *critique, questioning of assumptions*, and *praxis*, Wilson reflects on her role as an educator and provides insights into how CRP can be integrated into meaningful learning experiences. Learners are required to complete a reflexive critical skills blog and a critical essay but a range of formative activities are used to scaffold learning and demonstrate the links between sustainability theory and practice.

Schott adopts an entirely different approach to help learners link theory with practice by using virtual fields trips to immerse learners in a complex, three-dimensional learning space ("Digital Immersion for Sustainable Tourism Education: A Roadmap to Virtual Fieldtrips"). Students are able to explore a virtual 3D island setting in a developing country by accessing a range of embedded media, including images, video, audio, documents and spreadsheets. In Chap. 15 Tarrant et al. describe how learners develop humility, sustainability values and global citizenship while studying abroad under the Global Programs in Sustainability (GPS) initiative. The final contribution by Buijtendijk and van der Donk ("Sustainability in Tourism: A Corporate Perspective") examines the design of a 6-week intensive course developed to help learners understand the complexities of CSR. Learners complete a real-life consultancy assignment requiring them to work in international project teams to assess a 'client's' CSR activities in order to identify strategic improvements.

In summary, several important threads can be drawn together across the chapters that make up this volume. Firstly, EfS in tourism needs to be carefully integrated into the formal curriculum from early childhood education through to tertiary education. Secondly, several authors note that EfS is not just about sustainability knowledge, but also about practice, reflection and reflexivity. This requires learners to consider sustainability from multiple and often competing perspectives and disciplinary viewpoints. Thirdly, EfS in tourism involves life long and life wide learning that transcends time and place and extends beyond formal education settings to include stakeholders such as tourists, governments, businesses and communities to solve complex problems. Finally, effective EfS is underpinned by intrinsic values linked to social and environmental well-being.

17.3 Further Reflections on EfS in Tourism

The 16 chapters previous to this one represent the first and most comprehensive attempt to address EfS within the context of tourism practice and education. Despite the range of contributions and topics included there are some important topics that were not able to be covered either because appropriate contributors were not available or because the topic has not yet been given sufficient consideration to be discussed at length. This section acknowledges these major gaps in the coverage of the book and seeks to provide some initial thoughts on the nature and importance of these topics. Two main types of gaps can be identified—key stakeholders and key dimensions of EfS. In the first chapter Fig. 1.1. provided a framework for EfS in tourism that identified a range of key stakeholders or types of learner that needed to be considered in the development of effective EfS in tourism. An examination of these suggests that little consideration has been given in the present volume to learners in the technical and vocational education sector, elected or appointed government officials, or members of non-government organisations (NGOs) that get involved in tourism. Similarly, the dimensions of curriculum design and the role of technology have been identified as important but not considered in detail.

EfS in Tourism in the Technical and Vocational Sector

Discussions about EfS have not been restricted to the school or university sectors with considerable attention paid to the implementation of EfS principles and topics into technical and vocational education and training (TVET) (cf. Goldney, Murphy, Fien, & Kent, 2007; Willis, McKenzie, & Harris, 2008). Nor surprisingly there is considerable commonality between discussions of EfS in TVET and discussions of

EfS in other formal educational sectors. Fien and Maclean (2008), for example, provide lists of topics, skills and educational approaches that include things such as the need to understand global production and consumption systems, awareness of environmental concepts such as life-cycle analysis, inter and intra-generational equity, eco-efficiency, ecological footprints, and biodiversity, the development of respect for nature, care for community, respect for cultural diversity and gender equality, and support for participatory democracy, a focus on negotiation and conflict resolution skills, the ability to work cooperatively, creativity, flexibility, and participation in real world cooperative learning activities conducted with local communities. All of these are very similar to the lists provided in Chap. 1. Similarly, there is common ground in the discussion of barriers to the implementation of EfS in TVET (Arenas, 2008). The literature on EfS and TVET does, however, note some unique features of TVET including a strong focus on competency based learning and assessment and the importance of TVET in programs to improve social sustainability through education for employment for disadvantaged and marginalised groups (Arenas, 2008).

Despite recognition that tourism and hospitality are common areas of attention for TVET (Gough, 2008) there is no specific discussion in the wider TVET literature of EfS and tourism specific training. Much of the existing literature is about new sustainability focused jobs or on adaption in sectors such as manufacturing and construction (McDonald, 2014). The tourism literature is also generally silent on the links between sustainability and tourism vocational education. Chang and Hsu (2010) provide a typical discussion of vocational curriculum design for tourism that focusses on industry needs and practical skills with no consideration given to issues of sustainability. A notable exception to this is the work by Lewis (2005) where consultation with a wide range of community stakeholders from three Caribbean islands was used to develop a different approach to tourism TVET that incorporated more liberal education elements and advocated the inclusion of developing a holistic understanding of tourism and improving knowledge of social and environmental impacts into TVET programs. A similar approach is described for the Maldives by Shakeela (2008) who also questions the common practice of adopting western approaches to TVET. Further investigation into, and support for, the adoption of EfS in tourism in the TVET sector is clearly necessary.

EfS in Tourism for Government Officials and NGOs

Government officials and staff from NGOs share three common features. First, they are recognised as key stakeholders in tourism planning, development and management (Churugsa, McIntosh, & Simmons, 2007; Kennedy & Dornan, 2009). Second, they are often criticized for having limited knowledge and understanding of tourism (Harrison, 2008; Zahra, 2010). Thirdly, they are rarely discussed as potential learners or targets for tourism and/or sustainability education. A common theme in analyses of government officials and implementation of sustainable tourism policies is that their effectiveness is often significantly limited by a lack of understanding and experience of tourism with evidence from both different levels of government and across a wide range of locations including Africa (Slocum & Backman, 2011), the US (McGehee & Meng, 2006), New Zealand (Zahra, 2010), Thailand (Krutwaysho & Bramwell, 2010) and Australia (Ruhanen, 2013). A consistent finding across all these studies is that government officials, both elected and appointed, often focus only on the potential economic benefits of tourism and have little understanding of the potential for negative impacts, how destination residents and businesses could be involved in tourism or the tourism supply or value chain. This is not surprising as government officials come from a wide variety of educational backgrounds and are unlikely to have much experience of tourism other than if they have been tourists (Harrill, 2004). Ruhanen (2013) concluded by suggesting that the same educational strategies and tools suggested for improving destination community knowledge of tourism to support effective public participation in tourism planning should also be used for improving the knowledge and skills of public planners.

The case for NGOs is a similar one, although the prevailing discussion has been focused on the role of NGOs in ecotourism (Butcher, 2006), pro-poor tourism (Harrison, 2008) and community based or community benefit oriented programs (Simpson, 2008). NGOs can be classified into multiple types but of particular interest to the present discussion is the distinction between those that are tourism focused or tourism specific, such as volunteer travel organisations (Kennedy & Dornan, 2009), and those whose focus primarily on issues beyond tourism such as community development or conservation but that get involved in tourism as a tool to achieve their conservation or development objectives (Butcher, 2007).

Whilst recognising the positive contributions of NGOs, a number of researchers have raised concerns about the actions and outcomes of NGO directed or supported tourism activities including a lack of understanding of tourism and its relationship to sustainability and development (Butcher, 2006, 2007; Dixey, 2008; Harrison, 2008; Simpson, 2008). This lack of tourism planning and management capacity differs from that described for government officials in that NGOs are aware of and often driven to avoid negative environmental and cultural consequences of tourism, but typically lack an understanding of the economic and financial aspects of tourism. A lack of understanding of tourism markets, market demand and marketing and limited skills in business planning or assessing financial viability of tourism proposals have been noted as specific issues for NGOs, especially those who do not have a tourism specific focus (Butcher, 2006, 2007; Dixey, 2008; Harrison, 2008; Simpson, 2008). Despite this consensus that lack of tourism knowledge is an issue for the sustainable development of tourism, few have discussed options for addressing this issue. Strasdas, Corcoran, Petermann, Bushell & Eagles (2007) offer an example of a training program focused on ecotourism in protected areas that was designed for a range of stakeholders including NGOs, but generally this is an area of tourism education that has been given little attention.

Innovation in Curriculum and Pedagogy

The discussion so far has focused on key stakeholders not discussed by other chapters in this volume. Attention turns now to the key dimensions and processes that enable EfS. Although various chapters have highlighted the importance of curriculum design it is useful to draw together these contributions while reflecting more specifically on what we mean by curriculum and what the gaps are in curriculum design for sustainability. The origins of the term 'curriculum' can be traced back to the Latin word for race track or race course, but the concept has evolved to represent the content and skills that make up a course of study or syllabus (Prideaux, 2003). A more expansive view is that the curriculum is more than content and extends to the entire learning experience including content, delivery, assessment, values, standards, outcomes and evaluation methods (Kerr, 1968). While this conceptualisation is easily applied to formal education settings, in informal learning the curriculum might be thought of in terms of what is delivered, how and when it is delivered and what is learnt.

Prideaux (2003) identifies several dimensions that are relevant to both formal and informal learning settings:

- *Situation*: the background, abilities and experience of learners
- *Intents*: the aims, goals and outcomes
- *Content*: scope, sequence, related to aims, related to practice
- *Teaching and learning*: variety of methods, opportunity for self-direction, learning in real life settings
- *Assessment and feedback*: formative and summative feedback
- *Organisation*: blocks, units and timetables
- *Evaluation*: questionnaires, focus groups, participation

Prideaux's (2003) framework highlights the situational dimension of the curriculum. This is particularly relevant to EfS because learning differs from person to person, context to context, and culture to culture. Individuals bring different understandings and past experiences to learning settings, and the challenge for educators is to cultivate a sustainable mindset by helping learners understand and reflect on the complexities of sustainability (Gretzel, Davis, Bowser, Jiang, & Brown, 2014; Kagawa, 2007). However, there are a number of barriers to achieving this lofty ambition, including an overcrowded curriculum, staff and student resistance or uncertainty about sustainability, and institutional constraints (Wilson & Von der Heidt, 2013). There is a need for further research and development of curricula and pedagogies that encourage deep learning and the ability to think in complex, critical, systematic, holistic and interdisciplinary ways (Biggs, 1991; Thomas & Day, 2014). Curricula and pedagogies building on the constructivist learning paradigms of Piaget and Inhelder (1971) and Vygotsky (1978) offer many possibilities for moving beyond teacher-centered and content-focused approaches to sustainability.

Curriculum design is influenced by a range of factors that may result in the *intended* curriculum not always being the same as the curriculum that is *delivered* or *experienced* by learners. As Lund-Durlacher highlights in her Chap. 6, the values and behaviors of educators also form part of what might be described as the *hidden* curriculum. In this volume, both Lund-Durlacher and Wilson highlight the importance of sustainability educators 'practicing what they preach'. Several authors also discuss the need to ensure that sustainability principles and values are embedded across the entire curriculum. As Lund-Durlacher points out, this requires that learning is carefully scaffolded and demands a level of collaboration between educators that is rarely seen in formal educational institutions. These shortcomings highlight a need for further research about new ways to educate for sustainability, as well within a sustainable tourism context (Falk, Ballantyne, Packer, & Benckendorff, 2012).

Technology Enhanced Learning
While the relationship between technology and EfS may at first appear tangential, if not incompatible, there are many opportunities to harness technology to develop a better understanding of sustainable tourism across all of the stakeholders identified throughout this edited volume. Part V of this book included two examples of the use of technology to help learners move beyond sustainability knowledge to reflective practice. Jennings and Kachel described a blended learning approach that included online learning resources consisting of written text, visuals, audio recordings, links to related readings, video and digital media and personal calculators/indices related to sustainability. Schott adopted a similar approach but learning objects were instead embedded in an immersive 3D virtual learning environment. Both approaches illustrate a desire to move beyond the use of technology as a repository of information by creating more interactive learning experiences that encourage exploration and discovery. The real potential for technology enhanced learning in EfS lies in its ability to create authentic learning environments.

Authentic learning refers to teaching and learning approaches that immerse learners in realistic or simulated situations (Herrington, Reeves, & Oliver, 2007). Authentic learning is based on the premise that people learn better in context (Dewey, 1966). Context is extremely important in tourism sustainability because of its complexity and strong emphasis on practical knowledge and skills. Herrington, Reeves, and Oliver (2010) suggest nine design elements for authentic online learning environments: (1) an authentic context; (2) authentic tasks and activities; (3) access to expert performances; (4) multiple perspectives; (5) collaboration; (6) reflection; (7) articulation; (8) coaching and scaffolding; and (9) authentic assessment and feedback. These elements are compatible with EfS across a range of settings and stakeholder groups. Examples of technological developments that support authentic EfS in tourism include virtual learning environments (VLEs) and virtual field trips (VFTs), simulations, branching narratives, augmented reality and learning analytics.

VLEs include many of the design elements identified by Herrington et al. (2010). Dillenbourg, Schneider, and Synteta (2002) argue that VLEs move beyond content repositories because they turn spaces into places by encouraging social interaction and co-creation. The representation of content in VLEs can vary from text to images, video and 3D immersive worlds. The VFT described by Schott in the Chap. 14 provides a good example of an immersive 3D VLE while the tools and calculators included in the online module described by Jennings and Kachel in their Chap. 12 offers an different approach. Other approaches include the use of existing virtual worlds such as Second Life to develop an understanding of sustainability principles. VLEs offer a number of educational advantages over real settings. They are always open and available to learners, they can be accessed instantaneously from anywhere with an Internet connection, and content can be tailored to specific educational objectives (Eckelman, Lifset, Yessios, & Panko, 2011). They can also be applied in informal learning settings to help destination communities visualise the consequences of different development decisions.

Simulations offer another useful application of VLEs for EfS in tourism. Their strength lies in their ability to help learners model the consequences of decisions in a risk free environment, and to understand interactions between variables in complex systems (Fabricatore & López, 2012). In formal education settings sustainability simulations can facilitate critical reflection on the outcomes of operational, strategic and ethical decision-making. Educators at the University of Wollongong in Australia have developed an Interactive Dynamic Learning Environment (IDLE) to help learners to reflect on the complexity of corporate social responsibility and sustainability issues. However, there is also scope to use simulations beyond formal education settings to support visitor learning, community capacity building and government decision making. A related development is the use of gamification in formal education and tourist settings. While some simulations have a serious gaming element, other applications of gamification support EfS through inquiry-based learning, discovery and exploration (De Freitas, 2006; Dieleman & Huisingh, 2006; Metcalf, Kamarainen, Tutwiler, Grotzer, & Dede, 2011). For example, a gaming approach can be used to create quests requiring learners in formal and informal settings to discover clues that may reinforce sustainability messages. A gamification approach could also be used to reward and symbolize status and recognition for sustainable behaviors through the use of leaderboards, badges and levels (Kapp, 2012).

Further opportunities are afforded by developments in mobile technologies, which allow content to be personalized and contextualized based on the characteristics, preferences and location of learners (Benckendorff, Sheldon, & Fesenmaier, 2014). The *Google Field Trip* app is a good example of the potential of these technologies, particularly when combined with wearable devices and augmented reality applications.

The final comment in this discussion of technology enhanced learning is about the emerging field of learning analytics. Learning analytics involves the use of 'big data' to understand learning behavior and outcomes. In the context of EfS, there is scope to collect a range of analytics from mobile devices and VLEs to better understand how learners respond to various technology enhanced teaching and learning approaches. Siemens and Long (2011, p. 30) argue that learning analytics "can penetrate the fog of uncertainty around how to allocate resources, develop competitive advantages, and most important, improve the quality and value of the learning experience." While learning analytics are in an early stage of implementation and face a number of challenges, they offer a number of opportunities to enhance EfS not only in formal educational settings but also in visitor interpretation.

17.4 EfS: Moving Towards Sustainable Tourism Futures

One key conclusion that can be drawn from the various contributions to the present volume is that we need to change not only how we teach tourism, but what we teach about tourism. More specifically we need to more critically analyse how tourism affects all dimensions of individual and community well-being or quality of life and encourage alternative and more responsible approaches to tourism planning and management across the range of key stakeholders. Put more simply we have to determine, teach and practice what Jamal (2004) has called good tourism. We need to more explicitly and actively promote ethical and responsible approaches to tourism practice and we need to do more than describe such approaches in our teaching. As Shrivastava (2010, p. 442) notes, "my cognitive and intellectual understanding of sustainability issues did little to make me lead a more sustainable life". Simply describing sustainability issues and suggesting alternative tourism forms will not achieve sustainability. Understanding and adopting the principles and approaches suggested in the wider EfS literature may be a critical step towards achieving this goal.

References

Arenas, A. (2008). Connecting hand, mind, and community: Vocational education for social and environmental renewal. *The Teachers College Record, 110*(2), 377–404.

Benckendorff, P. J., Sheldon, P. J., & Fesenmaier, D. (2014). *Tourism information technology* (2nd ed.). Oxenford: CAB.

Biggs, J. B. (1991). *Teaching for learning: The view from cognitive psychology*. Hawthorne, VIC: ACER.

Bramwell, B., & Lane, B. (2013). Getting from here to there: Systems change, behavioural change and sustainable tourism. *Journal of Sustainable Tourism, 21*(1), 1–4.

Buckley, R. (2012). Sustainable tourism: Research and reality. *Annals of Tourism Research, 39*(2), 528–546.

Butcher, J. (2006). A response to building a decommodified research paradigm in tourism: The contribution of NGOs by Stephen Wearing, Matthew McDonald and Jess Ponting. Journal of Sustainable Tourism, 13(5), 2005, pp. 424–455. *Journal of Sustainable Tourism, 14*(3), 307–310.

Butcher, J. (2007). *Ecotourism, NGOs and development: A critical analysis*. Abingdon: Routledge.

Carp, R. M. (2012). Toward a resilient academy. In L. F. Johnston (Ed.), *Higher education for sustainability* (pp. 223–237). New York: Routledge.
Chang, T. Y., & Hsu, J. M. (2010). Development framework for tourism and hospitality in higher vocational education in Taiwan. *Journal of Hospitality, Leisure, Sport and Tourism Education, 9*(1), 101–109.
Churugsa, W., McIntosh, A. J., & Simmons, D. (2007). Sustainable tourism planning and development: Understanding the capacity of local government. *Leisure/Loisir, 31*(2), 453–473.
De Freitas, S. I. (2006). Using games and simulations for supporting learning. *Learning, Media and Technology, 31*(4), 343–358.
Dewey, J. (1966). *Democracy and education*. New York: Free Press.
Dieleman, H., & Huisingh, D. (2006). Games by which to learn and teach about sustainable development: Exploring the relevance of games and experiential learning for sustainability. *Journal of Cleaner Production, 14*(9), 837–847.
Dillenbourg, P., Schneider, D., & Synteta, P. (2002). Virtual learning environments. In *Proceedings of the 3rd Hellenic conference: 'Information & communication technologies in education'* (pp. 3–18).
Dixey, L. (2008). The unsustainability of community tourism donor projects: Lessons from Zambia. In A. Spenceley (Ed.), *Responsible tourism: Critical issues for conservation and development* (pp. 323–341). Abingdon: Earthscan.
Eckelman, M. J., Lifset, R. J., Yessios, I., & Panko, K. (2011). Teaching industrial ecology and environmental management in Second Life. *Journal of Cleaner Production, 19*(11), 1273–1278.
Fabricatore, C., & López, X. (2012). Sustainability learning through gaming: An exploratory study. *Electronic Journal of e-Learning, 10*(2), 209–222.
Falk, J. H., Ballantyne, R., Packer, J., & Benckendorff, P. (2012). Travel and learning: A neglected tourism research area. *Annals of Tourism Research, 39*(2), 908–927.
Fien, J., & Maclean, R. (2008). The legacy of the Bonn declaration. In J. Fien, R. Maclean, & M. G. Park (Eds.), *Work, learning and sustainable development: Opportunities and challenges* (Vol. 8, pp. xix–xxxv). Berlin: Springer.
Goldney, D., Murphy, T., Fien, J., & Kent, J. (2007). *Finding the common ground: Is there a place for sustainability education in VET?* National Centre for Vocational Education Research. http://files.eric.ed.gov/fulltext/ED499704.pdf. Accessed 12 Dec 2014.
Gössling, S., Hall, C. M., Ekström, F., Engeset, A. B., & Aall, C. (2012). Transition management: A tool for implementing sustainable tourism scenarios? *Journal of Sustainable Tourism, 20*(6), 899–916.
Gough, S. (2008). TVET as sustainable investment. In J. Fien, R. Maclean, & M. G. Park (Eds.), *Work, learning and sustainable development: Opportunities and challenges* (Vol. 8, pp. 107–115). Berlin: Springer.
Gretzel, U., Davis, E. B., Bowser, G., Jiang, J., & Brown, M. (2014). Creating global leaders with sustainability mindsets—Insights from the RMSSN Summer Academy. *Journal of Teaching in Travel & Tourism, 14*(2), 164–183.
Harrill, R. (2004). Residents' attitudes toward tourism development: A literature review with implications for tourism planning. *Journal of Planning Literature, 18*(3), 251–266.
Harrison, D. (2008). Pro-poor tourism: A critique. *Third World Quarterly, 29*(5), 851–868.
Herrington, J., Reeves, T., & Oliver, R. (2007). Immersive learning technologies: Realism and online authentic learning. *Journal of Computing in Higher Education, 19*(1), 80–99.
Herrington, J., Reeves, T., & Oliver, R. (2010). *A guide to authentic e-learning*. New York: Routledge. http://researchrepository.murdoch.edu.au/1903/.
Jamal, T. B. (2004). Virtue ethics and sustainable tourism pedagogy: Phronesis, principles and practice. *Journal of Sustainable Tourism, 12*(6), 530–545.
Kagawa, F. (2007). Dissonance in students' perceptions of sustainable development and sustainability: Implications for curriculum change. *International Journal of Sustainability in Higher Education, 8*(3), 317–338.

Kapp, K. M. (2012). *The gamification of learning and instruction: Game-based methods and strategies for training and education.* San Francisco: Wiley.

Kennedy, K., & Dornan, D. A. (2009). An overview: Tourism non-governmental organizations and poverty reduction in developing countries. *Asia Pacific Journal of Tourism Research, 14*(2), 183–200.

Kerr, L. (Ed.). (1968). *Changing the curriculum.* London: University of London Press.

Krutwaysho, O., & Bramwell, B. (2010). Tourism policy implementation and society. *Annals of Tourism Research, 37*(3), 670–691.

Lewis, A. (2005). Rationalising a tourism curriculum for sustainable tourism development in small island states: A stakeholder perspective. *Journal of Hospitality, Leisure, Sport and Tourism Education, 4*(2), 4–15.

McDonald, G. (2014). *Unpacking the green skills challenge.* Paper presented at Vision 2020 TAFE Directors Australia National Conference. http://www.tda.edu.au/cb_pages/files/Unpacking%20green%20skills_Final%20GMcD%20220811docx.pdf. Accessed 12 Dec 2014.

McGehee, N. G., & Meng, F. (2006). The politics of perception: Legislative images of the tourism industry in Virginia and North Carolina. *Journal of Travel Research, 44*(4), 368–378.

Metcalf, S., Kamarainen, A., Tutwiler, M. S., Grotzer, T., & Dede, C. (2011). Ecosystem science learning via multi-user virtual environments. *International Journal of Gaming and Computer-Mediated Simulations (IJGCMS), 3*(1), 86–90.

Piaget, J., & Inhelder, B. (1971). *Psychology of the child.* New York: Basic Books.

Prideaux, D. (2003). ABC of learning and teaching in medicine: Curriculum design. *BMJ: British Medical Journal, 326*(7383), 268–270.

Ruhanen, L. (2013). Local government: Facilitator or inhibitor of sustainable tourism development? *Journal of Sustainable Tourism, 21*(1), 80–98.

Shakeela, A. (2008). Tourism and hospitality education: The panacea for sustainable tourism development in the Maldives? In S. Richardson, L. Fredline, A. Patiar, & M. Ternel (Eds.), *CAUTHE 2008: Tourism and hospitality research, training and practice; "Where the 'bloody hell' are we?"* (pp. 578–587). http://search.informit.com.au/documentSummary;dn=994616040129951;res=IELBUS. Accessed 12 Dec 2014.

Shrivastava, P. (2010). Pedagogy of passion for sustainability. *Academy of Management Learning & Education, 9*(3), 443–455.

Siemens, G., & Long, P. (2011). Penetrating the fog: Analytics in learning and education. *Educause Review, 46*(5), 30–32.

Simpson, M. C. (2008). Community benefit tourism initiatives—A conceptual oxymoron? *Tourism Management, 29*(1), 1–18.

Slocum, S. L., & Backman, K. F. (2011). Understanding government capacity in tourism development as a poverty alleviation tool: A case study of Tanzanian policy-makers. *Tourism Planning & Development, 8*(3), 281–296.

Strasdas, W., Corcoran, B., Petermann, T., Bushell, R., & Eagles, P. (2007). Capacity-building for ecotourism: Training programmes for managers of protected areas. In R. Bushell & P. Eagles (Eds.), *Tourism and protected areas: Benefits beyond boundaries* (pp. 150–167). Wallingford: CABI.

Thomas, I., & Day, T. (2014). Sustainability capabilities, graduate capabilities, and Australian universities. *International Journal of Sustainability in Higher Education, 15*(2), 208–227.

Vygotsky, L. S. (1978). *Mind in society.* Cambridge, MA: Harvard University Press.

Willis, P., McKenzie, S., & Harris, R. (Eds.). (2008). *Rethinking work and learning: Adult and vocational education for social sustainability* (Vol. 9). Berlin: Springer.

Wilson, E., & Von der Heidt, T. (2013). Business as usual? Barriers to education for sustainability in the tourism curriculum. *Journal of Teaching in Travel & Tourism, 13*, 130–147.

Zahra, A. L. (2010). A historical analysis of tourism policy implementation by local government. *Journal of Tourism History, 2*(2), 83–98.

Index

G. Moscardo, P. Benckendorff (eds.), *Education for Sustainability in Tourism*, CSR, Sustainability, Ethics & Governance, DOI 10.1007/978-3-662-47470-9

MIX
Papier aus verantwortungsvollen Quellen
Paper from responsible sources
FSC® C105338

If you have any concerns about our products,
you can contact us on
ProductSafety@springernature.com

In case Publisher is established outside the EU,
the EU authorized representative is:
Springer Nature Customer Service Center GmbH
Europaplatz 3, 69115 Heidelberg, Germany

Printed by Libri Plureos GmbH
in Hamburg, Germany